PENALTIES FOR SHIP POWERING DUE TO UNFAVORABLE UNDERWATER HULL CONDITION

By

Sunil Sarangi

About Author

Ballast Water and Invasive Species issue is a subject of high importance, Sunil Sarangi is one of the Top experts in the world when it comes to ballast water treatment systems.

Sunil Sarangi began working on the Ballast water treatment systems projects in the United States. Sunil graduated from Birla Institute of Technology (BITS) with a Bachelor of Science in Marine Engineering. After obtaining his degree, he continued into graduate school to obtain his Master of Science in Ocean Engineering.

He also has an MBA with a specialization in Project Management.

During the past decade his work has focused upon developing cost-effective methods for reducing the spread of aquatic invasive species, the most widely referenced paper (Pimental et al. 2005) on this issue reports that invasive species cost the United States more than $120 billion in damages every year.

His work over the past 12 years can be summarized as covering the following subjects:

• the design and development of Ballast Water Treatment Systems.

• investigations into the effectiveness of techniques and technologies for measuring and reducing aquatic nuisance species.

• As a lead manager, successfully completed more than $40 million USD projects related to Ballast Water Treatment Systems.

Sunil Sarangi has authored several articles on Marine engineering and Ballast Water Treatment issues and has spoken at numerous symposia and conferences during the past decade.

His interest in the field led him to write articles that are published in Yahoo news and many other prominent sites.

He has founded Eco Marine, a Ballast Water Treatment Systems Design and consultants' firm that helps to protect and control aquatic invasive species.

Abstract

PENALTIES FOR SHIP POWERING DUE TO UNFAVORABLE UNDERWATER HULL CONDITION

This book presents the results of an investigation into the performance of ship's hull condition. Assessment of the fouling condition was carried out using visual observations and experimental techniques. Measurements of the paint condition were used to find potential impact on fuel penalties.

The investigation was carried out at three different dry-docks monitoring three different hull conditions of ships. Fuel consumption penalties due to the underwater condition of the hull significantly contribute to the operational cost of the ship.

The condition of hull coatings is crucial to the performance of ships. Frictional drag alone can account for as much as 90% of the total resistance required to move a vessel. In general, it has been shown that an increase in average hull roughness of 20 microns will result in a 1-2% increase in frictional drag, deteriorated coating and slime fouling may add 11-20% and the presence of calcareous fouling can increase drag by over 40%.

For these reasons, attention must be paid to the selection and maintenance of the submerged outer hull coating.

The main factors to be considered include performance, environmental controls, film thicknesses and the methods of application.

The experimental results are compared to values based on their coating condition, number of dry dockings, age of each ship, paint thickness and hull roughness conditions. The experimental results agree favorably.

Contents

List of Keywords

Anti-Fouling

Fuel Penalty

Dry dock

Hull Roughness Gauge

Dry Film Thickness (DFT)

Wet Film Thickness (WFT)

Proscope

Elcometer

Blisters

Biofouling

Microns

Blast Cleaning

Scraping

Sampling

Alligatoring

Touch dry

Heavy Fuel Oil

List of Figures

List of Tables

List of Abbreviations

MHR	Mean Hull Roughness
AHR	Average Hull Roughness
DMES	Department of Marine and Environmental Systems
ICCP	Impressed Current Cathodic Protection
DC	Direct Current
AFS	Antifouling System
IMO	International Maritime Organization
HFO	Heavy Fuel Oil
WFT	Wet Film Thickness
DFT	Dry Film Thickness
MDO	Marine Diesel Oil
D/D	Dry dock
SD	Standard Deviation

Acknowledgements

Many people have contributed to this project in more ways than hands-on assistance. Without these people, this project would have never been completed.

First, I would like to thank Dr. Geoffrey W.J. Swain for his invaluable support of my academic accomplishments and always teaching me more than I ever wanted to know about the safe practices of work and reminding me that things have to be done now not later.

I want to express my appreciation to the Royal Caribbean Cruise Lines for giving me an opportunity to visit different dry docks.

Finally, I would like to thank my friends and family who have been by my side throughout this process. The distractions, guidance, and support were all valued.

Dedication

This work is dedicated to two special people in my life.

To my dad, Vijaya Kumar, for his encouraging and support

and

To my mom, Vijaya Kumari, for her love and care

CHAPTER I

INTRODUCTION

1.0 General

Increases in the price of oil and regulations preventing the continued use of the tributyltin self-polishing copolymer antifouling paints (TBT_SPC) has forced the shipping industry to reassess the methods by which they can maintain their hulls in a smooth and fouling free condition. I had the opportunity to investigate the performance of underwater hull coatings on three ships: Grandeur of the Seas, Mariner of the Seas and Jewel of the Seas (figures 1-3) The study included observations and measurements made during three dry dockings, access to ship and hull coatings data, and the analyses of this data to develop guidelines for improved operations.

Figure 1 Grandeur of the Seas

Photo Courtesy www.cruisewarehouse.com/Royal%20Caribbean%20G

Figure 2 Mariner of the Seas

Photo Courtesy www.cruiseman.com/images/mairneroftheseas.jpg

Figure 3 Jewel of the Seas

Photo Courtesy www.cruiseserver.net/images/ships/rc_radiance.jpg

1.1 The Challenge

According to the International Maritime Organization (IMO) International Convention, after 1 January 2008 all vessels shall either: not bear anti-fouling containing TBT (tributyltin) on their hulls or external parts or surfaces; or apply a coating that forms a barrier (sealer) to such compounds leaching from the underlying non-compliant antifouling systems (International Maritime Organization).

In the 1980s and 90s the paint industries used tributyltin (TBT) as a biocide in anti-fouling paints to prevent fouling. The use of TBT systems provided long term fouling control with little or no settlement of fouling organisms (Jelic-Mrcelic *et al.* 2006). Unfortunately, this biocide is very harmful to certain marine organisms (Champ 2003), and regulations were enacted to restrict and prevent its use. The loss of TBT resulted in short term increases in hull roughness, fouling and fuel consumption (Abbott *et al.* 2000). This problem has been compounded by increases in the price of oil and concerns with respect to ship emissions

The two technologies that have replaced TBT are copper based biocide coatings and biocide-free silicone fouling release technologies. Recent research suggests that the fouling release coatings may outperform the biocide-based systems (Swain 2007) but unfortunately there is little ship hull data to corroborate these findings. Furthermore, it has been shown that even slime films can lead to significant increases in resistance and that heavy calcareous fouling may result in powering penalties of up to 86% [Schultz].

1.2 The Need for this Investigation

Fuel costs are increasing annually and this is driving the shipping companies to look for alternative options or to find the solutions to save the inefficient ways of wasting the fuel burn. Maintaining ship hulls and propellers in a smooth and fouling free condition is known to save large quantities of propulsive fuel, however, changes in antifouling coating technology, concerns about biocide output from antifouling paints, and changes in rules for the "Built for In-water Survey" (BIS) to increase the time between dry-dock from a 2.5 to 5-year schedule requires new approaches to selecting and managing antifouling systems. One initiative to reduce fuel consumption involves installing auxiliary diesel generators on gas turbine ships to produce electricity more efficiently and inexpensively. This effort can save up to $5 million per year, per ship.

1.3 Objectives

The objectives of this study were to obtain an understanding of the existing methods and technology used to maintain the ships in a smooth and fouling free condition and to use this knowledge for improved selection and management of antifouling coatings.

This was accomplished by attending the dry dockings of three ships, making a thorough visual inspection of the existing coatings, and monitoring the application of the new coatings. This book will summarize and discuss the findings.

CHAPTER II

BACKGROUND

2.1 Biofouling

Biofouling is the attachment and growth of living organisms such as algae, barnacles, mussels, and tube worms on manmade structures. If a ship is fouled it leads to roughness and reduced performance. This will compromise speed and maneuverability, increase fuel consumption, increase wear on machinery, transport invasive species and increase pollution.

In most instances, there is a sequence in the development of fouling on surfaces. It usually starts with organic films, bacteria and diatoms which form slime films (Figure 4). These are followed by larger fouling organisms. The development of fouling leads to penalties in ship performance, and, therefore, it needs to be removed mechanically by inwater hull cleaning or by regular scheduled dry dockings. Regular hull cleaning prevents calcareous fouling damaging the paint coatings and commercial and naval ships have demonstrated that savings in energy are possible by maintaining the hull and propeller surfaces in a smooth and fouling free condition [Naval Sea Systems]. More than 15% fuel savings can be obtained by cleaning of hull and polishing the propeller surfaces.

	Event	Illustration	Size, mm	Penalty
	Organic Film			?
	Bacterial Slimes		0.001 to 0.1	1-2%
	Diatoms & Filamentous Algae	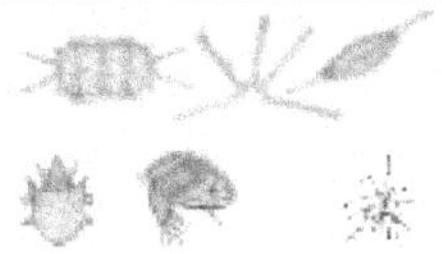	0.1 to 1.0	10-20%
	Barnacles, Bryozoans, Tubeworms, Macroalgae	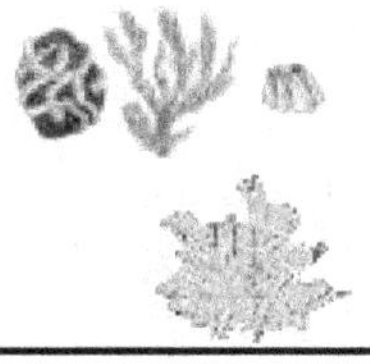	1.0 to 30	60+%
	Mussels, Oysters, Sponges	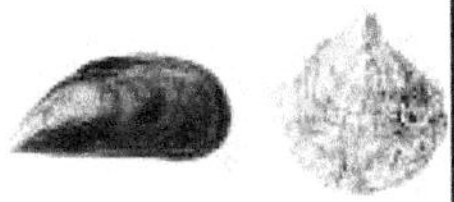	30 to 50	?

Figure 4 Different Stages of Fouling

2.1.1 Types of Fouling

Fouling is typically divided into four main types:
1. Slime, which includes bacteria and diatoms.
2. Weed Fouling, which includes green and red algae
3. Soft Fouling, which includes hydroids, bryozoans and sea squirts.
4. Shell Fouling, which includes barnacles, tubeworms and mollusks.

2.1.2 Where it Occurs

The majority of the fouling occurs when the ship is in port. This is due to an abundance of fouling settlement stages and low water movement. Fouling also tends to be more severe near the water line since plant fouling needs light (Yule and Walker, 1984). The boottop area is particularly vulnerable to slime and weed fouling, which is particularly unattractive. The area under the flat bottoms is usually clean of weed due to lack of light, however, it may become fouled by hard foulers such as barnacles and tubeworms.

2.1.3 How it Occurs

Fouling occurs by the arrival of juvenile stages of organisms that are present in the water to the ship hull surface. The presence and abundance of specific types of organisms is determined by the geographic location, the time of year and local conditions. If the organism finds the surface attractive for settlement, then it will attach itself to the surface. There are several different mechanisms by which the organisms adhere to surfaces and include:
- Glue method
- Stefan adhesion method
- Mechanical method
- Pressure difference method

Most organisms use glue to stick to a surface and they are very hard to remove (Denny, 1988). Slime consists of bacteria, protozoa, fungi and algae, and may attach to a surface within one hour of immersion to sea water (Navy Ship Technical Manual). The barnacle adheres to the surface during the cyprid stage where, once a suitable place is found, then it attaches temporarily using suction cups and Stefans adhesion. The strength of it becomes permanent by secreting glue (Singer,1968). After settlement adhesion can increase by over 5.5 times (Yule and Walker, 1984). Barnacles are the most abundant fouling seen in sea chests and stabilizer pockets. Advanced stages of fouling include barnacles, tubeworms, bivalves such as oysters, mussels, and hydroids such as coral or anemones.

2.2 Ship Hull Coatings

The purpose of a ship hull coating system is to protect the steel from corrosion and to provide an outer layer that will remain free from biofouling and in a smooth condition.

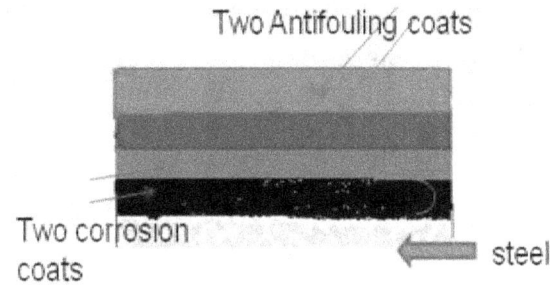

Figure 5 Typical Coating Schemes for SPC Paint

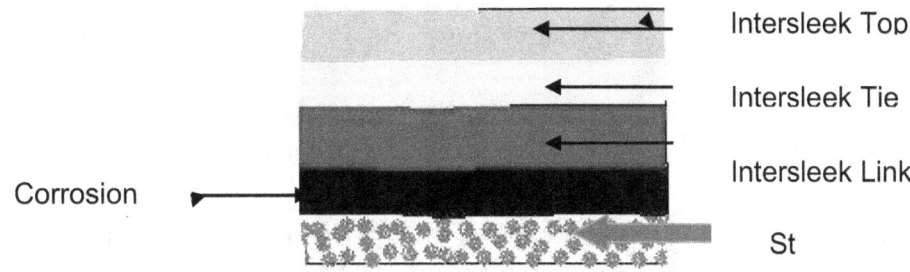

Figure 6 Typical Coating Schemes for Intersleek Paint

The typical coating system consists of a corrosion coat (usually epoxy) overcoated by an antifouling system (copper or silicone based) (Figures 5 and 6). The epoxy coatings work as a barrier to stop the oxygen and sea water ions from reaching the steel. A good quality barrier coat requires correct surface preparation of the steel and a minimum thickness of about 350 microns' dry film thickness (DFT). This usually requires two coats of about 150-200 microns DFT. The epoxy must adhere well to the steel and also provide a tie coat to which an antifouling top coat may be applied. Antifouling coatings are typically between 150- 200 microns DFT. The number of coatings applied will typically depending on the ship operation and design life. Copper based systems are typically applied at thicknesses ranging from 150 to 450 microns DFT. Silicone based systems are usually between 200 to 300 microns DFT.

2.2.1 Present Day Antifouling Paints

Antifouling paints may be classified as biocide based or biocide free (Table 1). Those containing biocides are classed as biocidal products and in many countries are regulated like pesticides. Most manufacturers removed TBT antifouling paint in 2004 after the IMO Antifouling Systems Convention was enforced (Antifouling Legislative Position, Feb, 2004). Today's biocide-based coatings rely on copper as the principal agent and this may be reinforced by zinc oxide and cobiocides such as zinc and copper pyrithiones, irgarol and isothiazolin. Alternatives to the biocide coatings are the silicone fouling release systems or hard cleanable coatings.

Table 1 Comparisons of Latest Coating Technology

Ablative, Eroding & Hybrid Coatings	Self-Polishing Coatings	Silicone Fouling Release Coatings	Cleanable Hard Coating
Biocides, Copper plus cobiocides	Biocides, Copper plus cobiocides	Biocide-free	Biocide-free
Soluble matrix rosin	Acrylic polymerized to form copper acrylate, silylacrylate, zinc acrylate	May become fouled	Will become fouled
Reinforcing polymer, acrylics, polyester, polyamide, vinyl	Soluble microlayer that polishes away	Fouling easily removed	Requires frequent and intensive cleaning
Thick leached layer	Thin leached layer	Smooth inert top coat	Smooth inert top coat
Added roughness	Self-smoothing	Coatings are soft	Coatings are hard
Service life depends on coating thickness and operation schedules	Prolonged service life	Long life expectancy	Long life
Sea water penetration	Polishing action determined by sea water conditions.	Low skin friction	Easy to apply
May requires in water cleaning	No in water scrubbing	Expensive	Inexpensive
Reduced biocide output	Controlled biocide output	Application issues	Invasive species issues

The application of antifouling coatings to the underwater hull is the most frequent work undertaken in the dry docks throughout the world. Once fouling is established on a ship hull it is hard to remove, and prevention is always better than cure. The choice of the underwater hull coating and long term maintenance is very important for the economy of the vessel. So the painting on the ship is critical.

The application of the paint is also expensive and requires attention to cleaning, surface preparation and the paint application. The higher the weight percent solids the higher the dry film thickness and this can help reduce application costs. Airless spray is the best proven method for application; however, losses may be between 20-40% during application. The weather condition plays a major role during painting and wind is the main factor. Painting should always be done in dry weather, and when possible moist conditions should be avoided. Thinning the paint should be discouraged as it will lead to inferior paint because the amount of solids will decrease which will reduce the resulting DFT and there is increased likelihood of entrained solvents.

2.2.2 Blistering

One of the major problems with the underwater hull coatings is blistering. Blistering is very hard to control. Blisters may be caused by rust, concentrations of chemical or cathodic protection. During the service of the ship, the blisters often burst and cause damage to the coating. During the dry dock, if the blisters are not punctured, then there is a possibility that oxygen may pass inside the blisters and cause corrosion of the underlying steel. During inspection blisters may be hard to find under the flats of the ship because of the reduced light intensity. If blisters are seen during the inspection, they should be marked for repair work.

2.3 Cathodic Protection

Cathodic protection of the underwater portion of a ship's hull is an essential part of corrosion control. This may be achieved by the use of sacrificial anodes (zinc or aluminum) or by an impressed current cathodic protection system (ICCP) using inert anodes and an external DC current source (figures 7). ICCP systems usually consist of four circular anodes, two placed forward and two placed aft. A daily log of the operating parameters is kept to include: ship's position, mean sea temperature, engine rpm, reference potential (port and starboard and fore and aft), anode current output and anode driving potential.

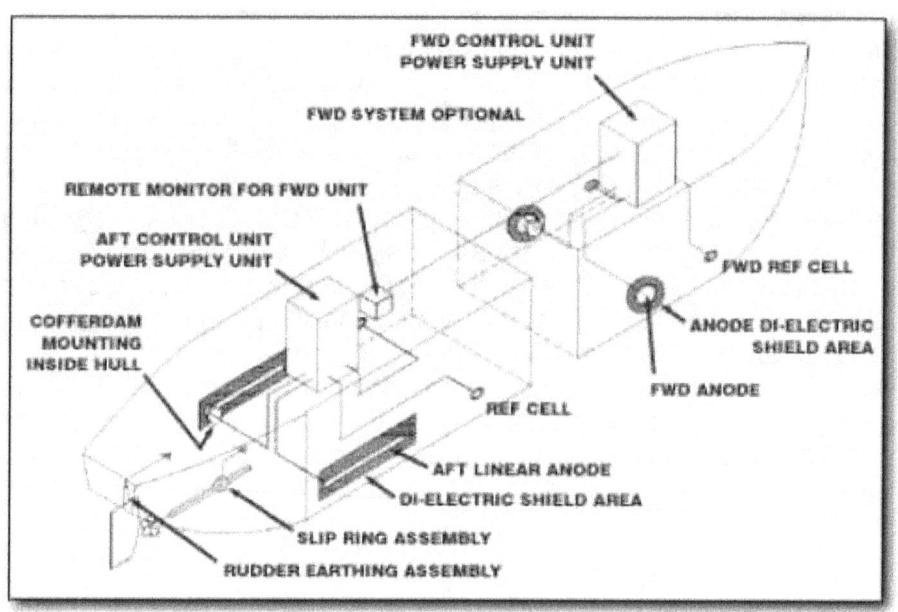

Figure 7 Typical Ship ICCP Layout (Photo courtesy of Cathelco UK)

Cathodic protection may cause the formation of blisters. This is the result of electro osmosis when the driving cathodic potential exceeds the strength of the coating. The difference in potential between the anode and the local cathode drives ions into the coating and causes blisters to form. This is more likely to occur when there is a high cathodic driving potential, the coating is thin or damaged, the coating is unsuitable to be used in conjunction with ICCP, or the coating is badly applied. Osmosis and electro osmosis tend to occur early in the lifetime of a coating while it retains a degree of plasticity.

If the system gets damaged or malfunctions the electrodes in ICCP draw very high currents producing hydroxyl ions at the cathode and chlorine at the anodes Care should be taken to not paint the anodes and reference cells while in dry dock.

2.4 Hull Roughness

Hull roughness is defined as the maximum height of peak to trough for a given length of 50 mm and is always expressed in microns. There are two types of underwater roughness
1. Physical roughness which include micro and macroroughness
   a. Microroughness is the paint surface profile.
   b. Macroroughness includes welds, mechanical damage and paint build up.
2. Biological roughness includes micro and macro fouling
   a. Microroughness includes the bacteria and diatom slime films
   b. Macroroughness includes weed, barnacles and tube worms.

22

2.5 Penalties for Ship Powering due to Hull Condition

Hull roughness and fouling contribute significantly to the viscous drag experienced by ships. It is generally accepted that "each 20-micron increase in hull roughness can result in a 1-2 % increase in fuel consumption" [Swain and Schultz] or a reduction in ship speed. It has been shown that hull roughness increases with the age of the ship due to breakdown of coating and added paint films at dry dockings (Akzo Nobel 2004) Ship operational conditions like time in port / at sea, geographical area, cleaning cycles, corrosion, interaction with the cathodic protection system, old paint, weld seams, and damage from mooring and anchoring will all contribute to the roughness. Torben Munk (Motorship Propulsion Conference 2006) suggests that added resistance due to fouling of hull and propeller varies from 6 to 80% and that over 50% of the world fleet has added resistance over 50%. Slime films up to 20% increase and weed and shell up to 60% increase.

2.6 Process of Dry dock

As soon as the vessel is dry docked and secured, the ships representative takes a walk along the length of the ship to observe the condition of the ship. This includes fouling and any obvious defects. The first thing to be done is to wash the entire hull of the ship with high pressure water (3,000 psi). This procedure removes most of the slime and weed; however, calcareous fouling may need mechanical removal. Once the hull is fully washed, the condition of the hull, stabilizers, rudders, propellers, (azipods), sea chests, bow and stern thrusters, reference electrodes and sacrificial and impressed current anodes is determined and decisions are made as to the work schedule. This includes the use of ultra-high-pressure water, slurry or sand blast, and grinding of the damaged areas. More detailed examination may also be made to include dry film thickness and hull roughness measurements. This provides a baseline for quality control. Once the hull is fully washed it is good to walk the length of the ship with a scraper in hand to look for the blisters and crack the blisters and mark the spots with a piece of chalk so that the blistered area can be ground back to a clean surface and can be painted. If blast cleaning is to be done, then it is best to chip off or scrape all the loose paint and hard fouling, so time and the amount of sand consumption will be less. Any oil contamination near the bow thrusters, and azipods, needs to be removed. If blisters are present, they should be removed as they can cause trouble in the future if they burst after the painting is done.

After the surfaces have been deemed suitable, the coating application starts. Close monitoring of paint application includes the quality of mixing of the paint, conditions in the dry dock (temperature/wind), WFT measurements during application to ensure correct DFT, and time between coatings. Freshly applied paint shall be protected from any other activity, for example no anchor chain maneuvers are to be allowed between the last low pressure fresh water rinse until the last coat of antifouling has cured to at least "touch dry". Ensure all scupper plugs are in good condition and not leaking during painting. If the painting needs to be done during the night, then appropriate lighting has to be installed.

When there is damage of the ICCP dielectric shield, the coating needs to be disked and replaced. This is accomplished by the application of a special epoxy compound (total DFT 800 microns) with a spatula immediately after blasting.

2.6.1 Preparation Grades-Blast Cleaning

There are several different standards that define the surface preparation of the hull (NACE, SSPC, and SA). The degree of surface preparation depends on the recommended practice required for the paint. For example, The Swedish standards are as follows:

- Sa 0: No preparation of surface.
- Sa 1: Light blast cleaning. The hull is washed for loose mill scale, rust, and foreign matter removal.
- Sa 2: Thorough blast cleaning. The surface shall be free from visible oil, grease and dirt, and from most of the mill scale, rust, paint and foreign matter.
- Sa 2 ½: Very thorough blast-cleaning. Mill scale, rust, and foreign matter shall be removed to such an extent that remnants only appear as slight stains on the surface.
- Sa 3: Blast cleaning to steel. The hull is pressure washed until all mill scale, rust and foreign matter is completely removed and the shining steel is visible.
- St 2: Thorough hand and power tool cleaning. Loose scale, rust and foreign matter should be removed.
- St 3: Extremely thorough scraping and wire brushing, disc-sanding, power-brushing, etc. After removing the dust, the surface looks like a metallic sheen.

CHAPTER III

METHODS

The following methods were developed to investigate the condition of the underwater portion of the ship's hull during dry docking. The methods were to provide data on the condition of the ship when she entered dry dock, during wash down and surface preparation, and after recoating.

3.1 Overall Procedures

The submerged portion of the hull was divided into 6 main areas: bulbous bow, waterline, sides, flats, keel and stern. The stabilizers, rudders, and propellers were inspected separately. The following methods and procedures were used for inspection and monitoring:

- Visual observations of the hull were made with the aid of video and still photography. This was completed for the following conditions:
  - immediately after draw down to observe biofouling
  - after high pressure washing to observe coating condition
  - during and after surface preparation
  - during and after coating application.
- Dry film thickness measurements were made after high pressure washing and repainting.
- Destructive coating thickness measurements (ASTM D4138-07) were made after high pressure washing.
- Wet film thickness measurements were made during coating application.
- Hull roughness measurements were made after high pressure washing and after coating application.

3.2 Method for Observing Hull Condition

Visual observations of the hull were made with the aid of video and still photography (Figure 8). This was completed immediately after draw down to observe biofouling, after high pressure washing to observe coating condition and during and after surface preparation and coating application.

Figure 8 Visual Observations of Paint Condition

3.2.1 Method for Rating Hull Fouling

Fouling is the term used to describe the growth of marine plants and animals on ships and structures. During the dry dock inspection, the types of organism present on the hull were identified and the percent area of the surface covered by the fouling types estimated. Fouling is multilayer and therefore when the individual percentages are added together, they may be greater than 100%.

The following types of organism were identified as occurring on the three ships inspected during work (Table 2 and Figures 9-12)

3.2.1.1 Types and descriptions of major fouling groups

Table 2 Types of Major Fouling Groups

Group	Description
Slime	Diatoms, initial algal germination, and low form algae, bacterial growth
Algae or Weed	Green or red mats of filamentous plants (e.g. *Ulva* spp.*, Ectocarpus*,)
Hydrozoan	Low form, highly branching organism
Encrusting Bryozoans	Hard colonial organisms with a low form, encrusting growth form. Up to several millimeters thick
Barnacles	Hard shelled crustacean that cements permanently to a surface. Whitish outer shell with tear-drop shaped orifice. Up to 30mm diameter and height
Tubeworms	Worms that form a hard calcareous or soft tube that is cemented to the surface

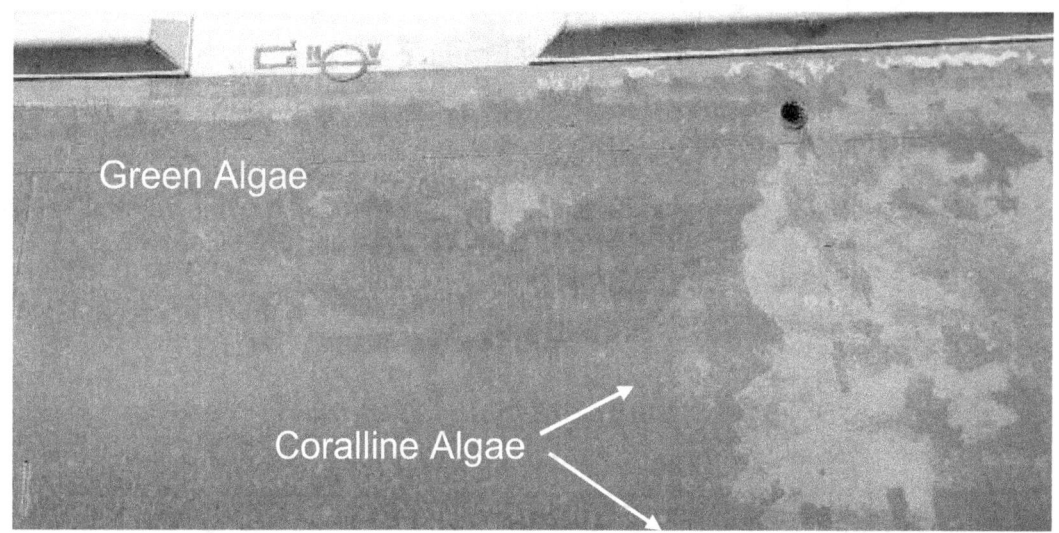

Figure 9 Slime Films, Green and Coralline Algae

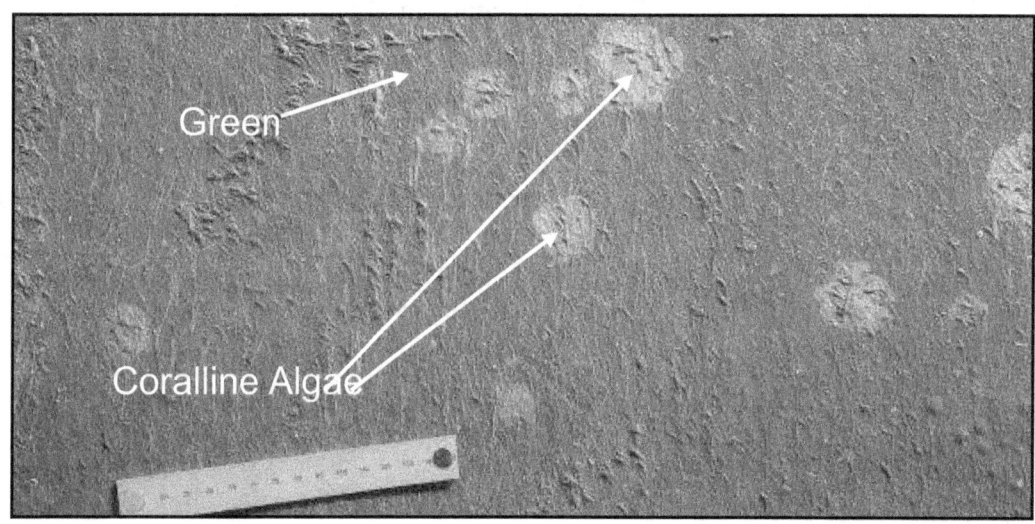

Figure 10 Slime Films, Green and Coralline Algae

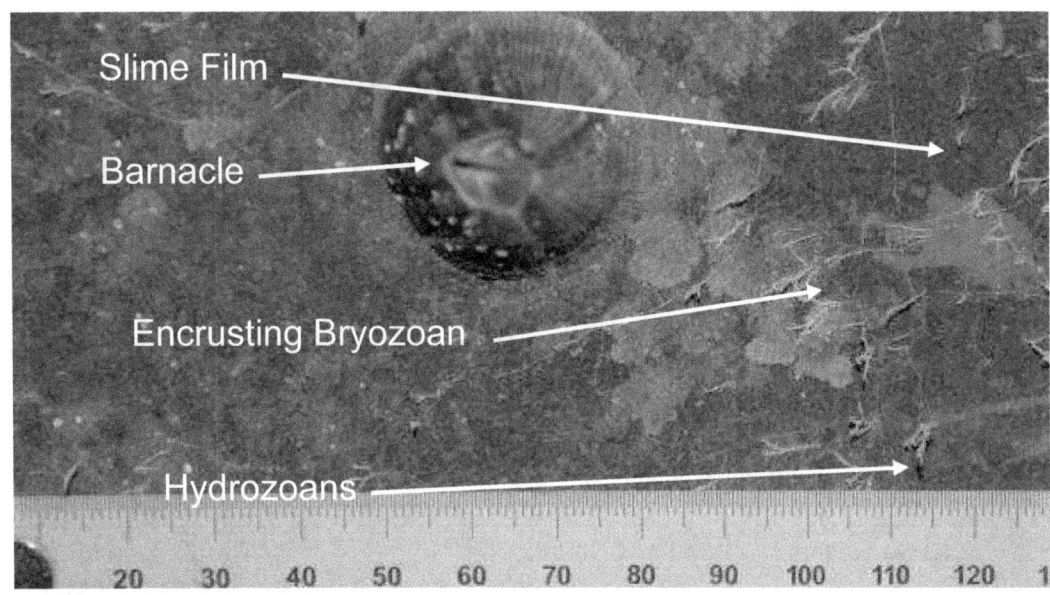

Figure 11 Slime Films, Encrusting Bryozoa, Hydrozoans and Barnacle

Figure 12 Slime Films, Encrusting Bryozoa, Barnacle and Tube Worms

3.2.1.2 US Navy Fouling Rating Method

The US Navy has developed a pictorial fouling rating for their ships (Figure 13) and recent research has related the fouling rating to the increase in powering required to maintain ship speed (Table 3). The US Navy fouling ratings and the percent cover by the major groups of fouling present were estimated for the major areas of the ship.

Table 3 US Navy Fouling Rating Method and predictions for change in shaft horse power to maintain speed

M.P. Schultz Effects of coating roughness and biofouling on ship resistance and powering. Biofouling, 2007; 23 (5/6): 331-341

NSTM Rating	Description	Predictions for change in SHP to maintain FFG Frigate at 15 knots
0	Hydraulically smooth surface	0%
0	Typical as applied antifouling coating	2%
10 - 20	Deteriorated coating or light slime	11%
30	Heavy slime	21%
40 – 60	Small calcareous fouling or weed	35%
70 – 80	Medium calcareous fouling	54%
90 - 100	Heavy calcareous fouling	86%

US Navy Hull Fouling Rating system

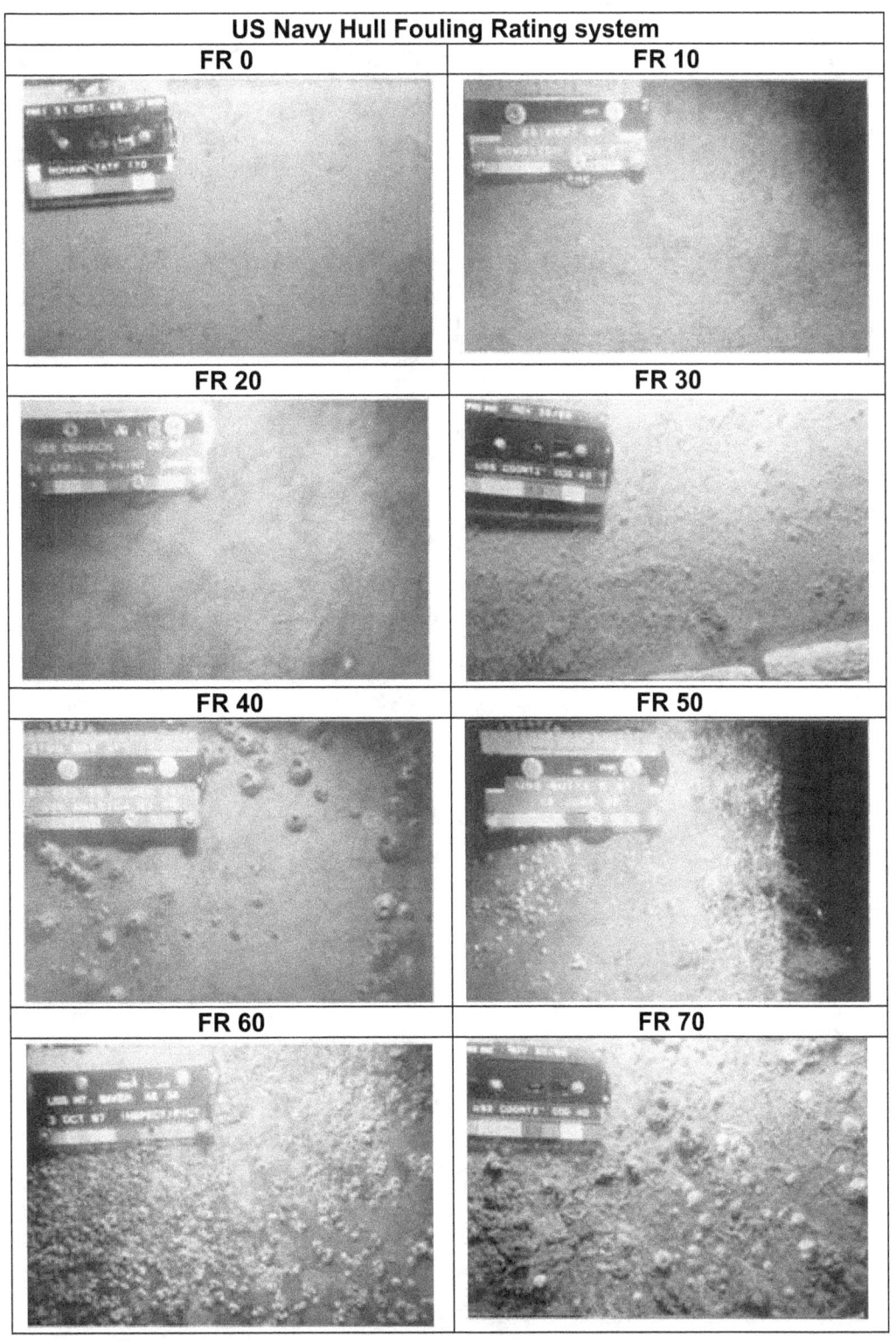

FR 0	FR 10
FR 20	FR 30
FR 40	FR 50
FR 60	FR 70

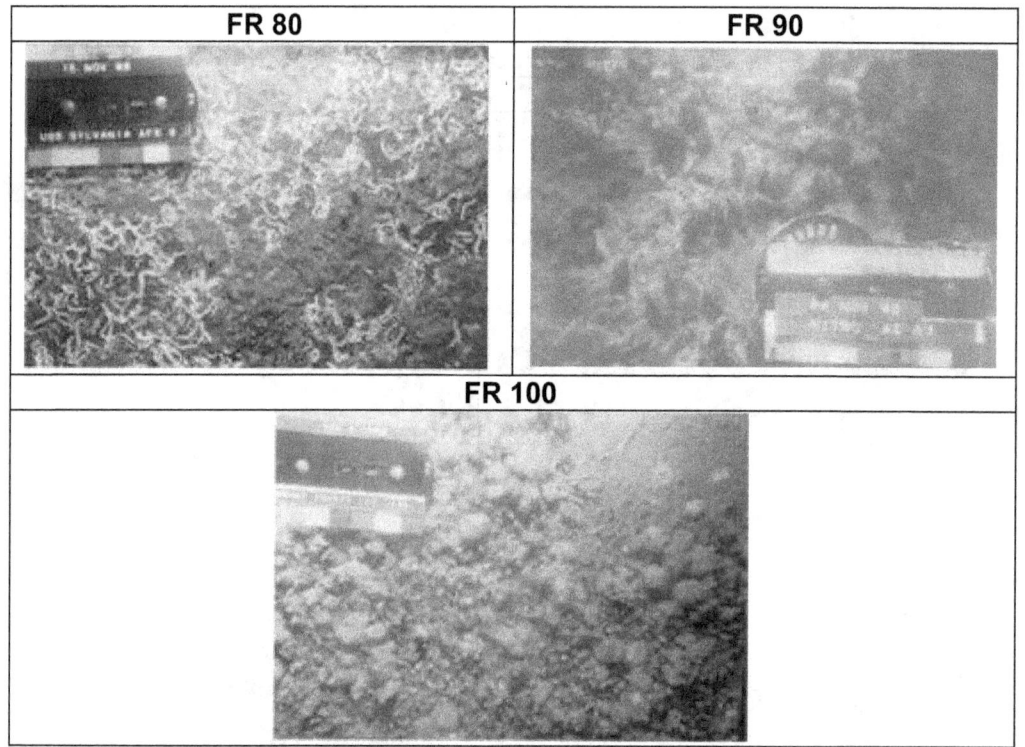

Figure 13 US Navy Foul Rating System

3.3 Method for Measuring the Coating Condition

Visual observations of coating condition were made after the high-pressure wash. The coating was categorized as good, orange peel, runs and sags, alligatoring, mechanical damage, blistering and peeling of the paint (Table 4).

Table 4 Types of Paint Defects and Descriptions

Good	Coating forming an intact and adhered smooth film of adequate thickness over the steel
Orange Peel	Dimpled orange skin like surface due to poor paint application
Runs and Sags	Uneven surface in the form of hills, valleys and beading due to too greater coating thickness during application
Alligatoring	Alligatoring is caused due to the paint solvent instability
Blistering	Round hemispherical projections of the coating away from the surface filled with fluid. Possibly caused by contamination of the steel surface (flash rusting) prior to coating application and interaction with impressed current cathodic protection
Peeling	Adhesion related failure of the coating to the steel probably caused by poor substrate condition during application
Mechanical Damage	Coating compromised by mechanical damage caused by anchor chain and fendering

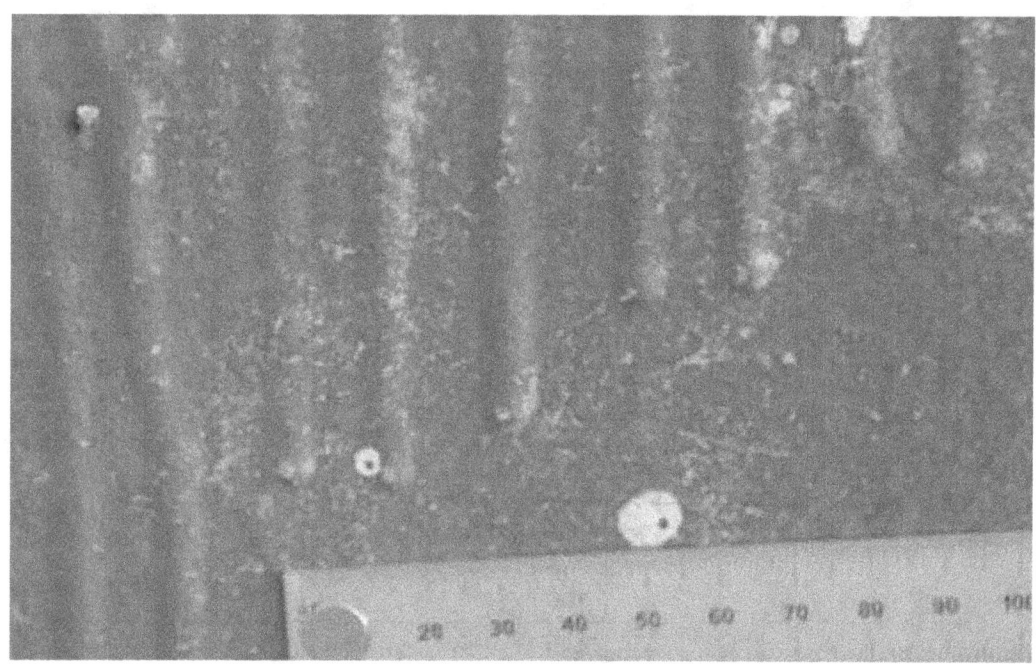

Figure 14 Runs and Sags of a Coating

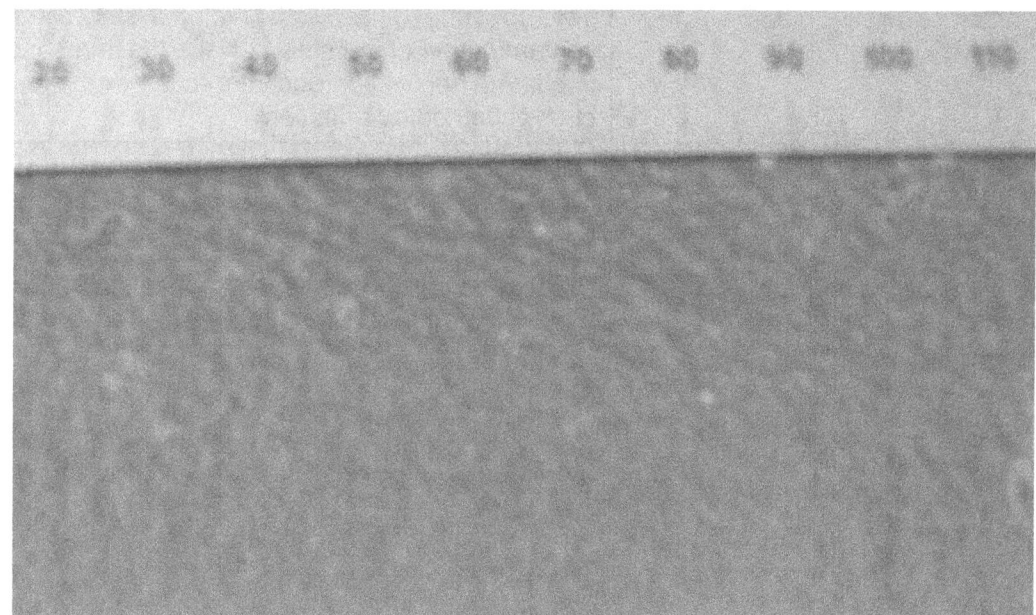

Figure 15 Slight Orange Peel

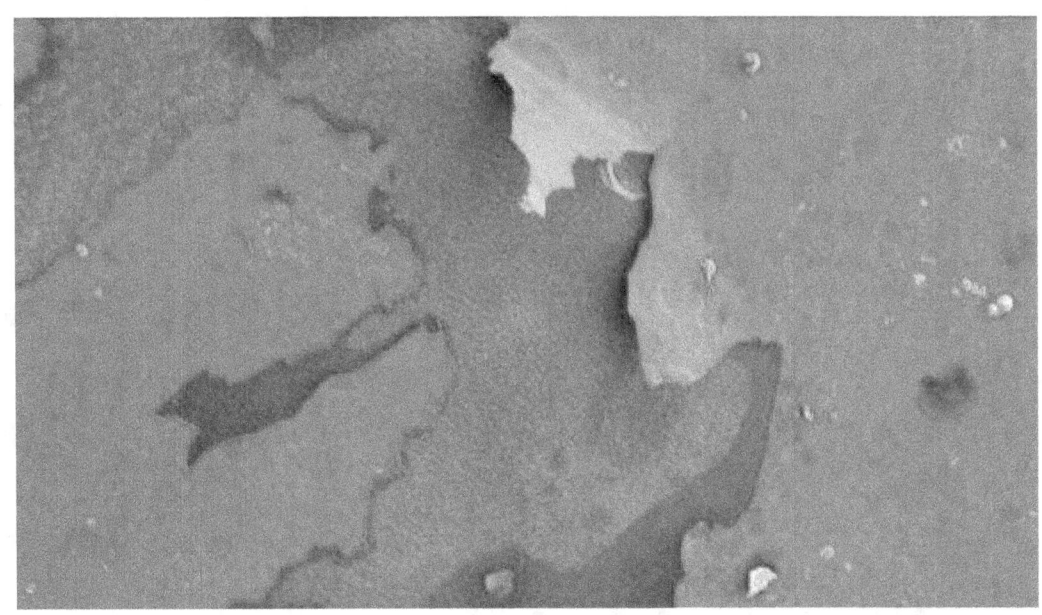

Figure 16 Peeling of Paint

Figure 17 Blisters

Figure 18 Mechanical Damage

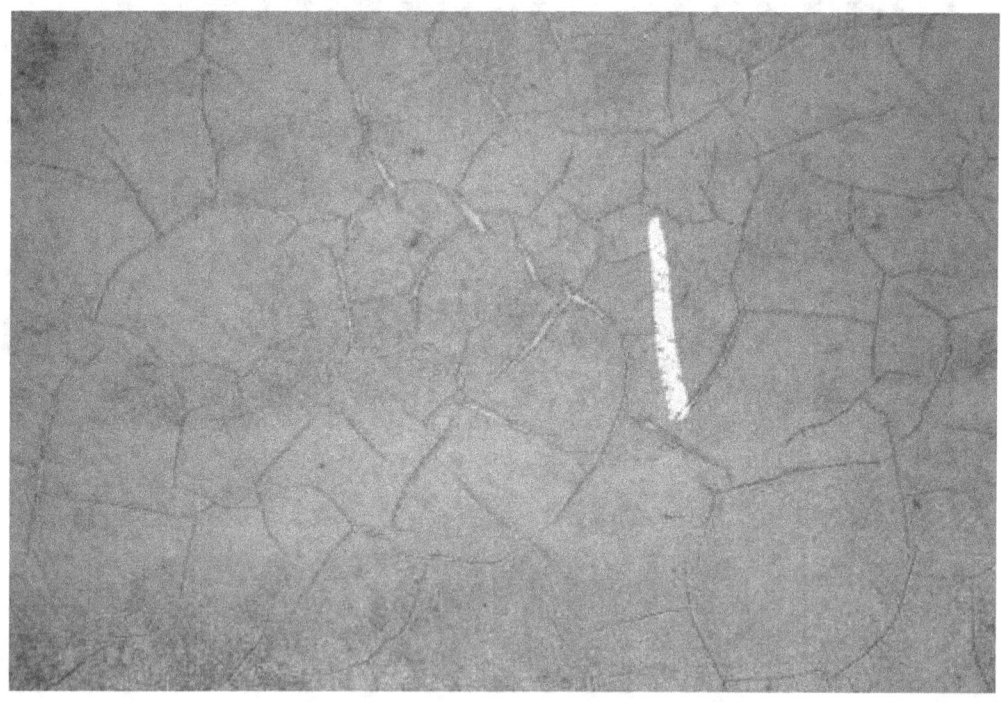

Figure 19 Alligatoring of Paint

3.4 Method for Measuring Paint Film Thickness

Three methods were used to measure paint film thickness. Two methods measured dry film thickness and one measured wet film thickness.

3.4.1 Dry Film Thickness (Elcometer)

Dry film thickness (DFT) of the coatings was measured using an Elcometer 345F Digital Thickness Gauge™. (Figure 20) The Elcometer 345F measures the thickness of non-magnetic coatings on steel or iron. It can be used on paint, plastic, galvanizing, enamel, hard chrome, and other coatings such as electroless nickel. The Elcometer 345 standard and top versions have built in statistics, and the top version has added memory for up to 10,000 readings. The probe of the instrument is placed on the test surface and set in place. The display will blink and then indicate the thickness of the coating. If the dotted line appears, the value is just greater than the range of the instrument, or the instrument is out of calibration.

Figure 20 DFT Measurements Using Elcometer

3.4.2 Dry Film Thickness (Destructive ASTM D4138-07)

This method measured the coating dry film thickness by microscopic observation of precision angular cuts in the coating film. A cutting tool with known angle was used to cut a groove through the coating to the bare steel (Figure 21). A Proscope™ digital imaging system was used to obtain a magnified image of the groove. This enabled coating layers and thickness to be calculated by multiplying the horizontal measurement by tan Ø.

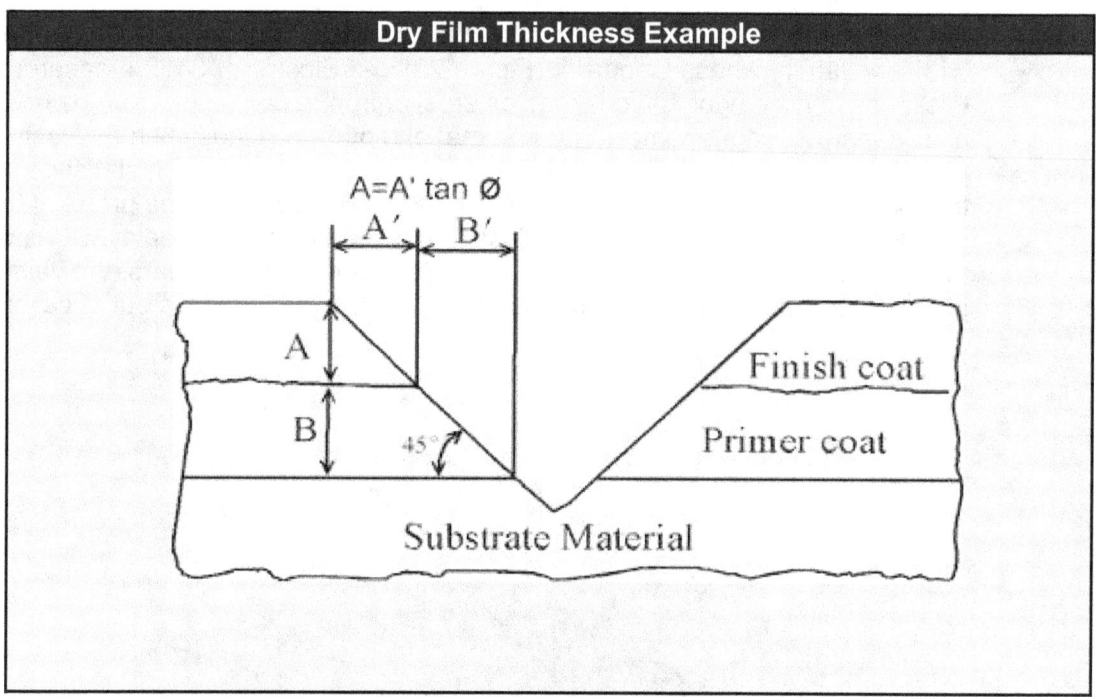

Figure 21 Coating Systems Cross Section Cut

3.4.3 Method for Measuring Wet Film Thickness.

Wet film measurements were made during coating application (Figure 22). The wet film thickness gauge is pressed into the wet coating and then takes the gauge back vertically and finds the deepest tooth of the gauge having paint on it and the immediate higher tooth that is not coated with paint. The true wet film thickness lies between these two readings.

Figure 22 Wet Film Thickness Gauge

3.5 Method for Measuring Hull Roughness

Hull roughness measurements were made after high pressure washing and also after coating application. Ship hull roughness was measured using the British Maritime Technology Hull Roughness Analyzer™ (Figure 23). This instrument measures the maximum peak to minimum valley height (Rt_{50}) over a 50 mm sample length over the surface. The surface probe was manually run over a distance of 750-1000 mm to generate 13-15 Rt_{50} readings, the average of which is the mean hull roughness (MHR) at that location. At least 100 locations distributed around the hull (including sides and flats) are required for statistical validity. The average of all these mean hull roughness gives the Average Hull Roughness (AHR).

The Measuring Head should be traversed parallel to the keel or, if known, the direction of the water flow, and should not be used on concave surfaces with a radius of less than 2.5 meters.

It is not always possible to obtain the desired number and distribution of measurements during the dry docking due to some limitations of time and manpower.

Figure 23 Hull Roughness Measurement Using Roughness Gauge

CHAPTER IV

RESULTS AND DISCUSSION

4.1 Introduction

Observations and data obtained for the three ships demonstrated that there were distinct differences in biofouling, coating condition, dry film thickness and hull roughness. These differences were reflected by the age of the ship, number of dry dockings, surface preparation, and coating application. This results and discussion are presented with respect to the ship characteristics, dry dock data and a discussion of the cost and selection of coating.

The ship characteristics includes: age of ship, existing hull coatings, operational area, speed, fuel consumption, hull cleaning, and previous dry dock schedules

The dry dock data includes: fouling condition, coating condition, paint thickness, hull roughness, dry dock procedures, how the ship left the dock.

Finally, coating selection and costs are discussed.

4.2 Characteristics of the Three Ships

The ships represent three different classes with the Mariner being the largest and Grandeur the smallest. Both Grandeur and Mariner were built at Kvaerner Masa yards in Finland and Jewel was built at Meyer Weft, Papenburg. In 2001 Grandeur of the Seas was refurbished.

The average speed for the ships is about 17 knots. Grandeur and Mariner use diesel electric engines for propulsion and Jewel uses gas turbines. There is a large difference b between the price for heavy fuel oil and marine diesel oil, and this increases the cost of fuel for Jewel. The antifouling paint systems differed for the three ships. Grandeur had several systems the most recent being Interspeed 640 tin free paint. Mariner and Jewel both had copper self-polishing copolymer systems.

4.2.1 Age of the Ship and Existing Hull Coatings

The condition of the ship hull and coatings is usually determined by the age of the ship. It is generally accepted that, as the age of the ship increases, hull roughness increases resulting in the increased power required to maintain the vessel speed [Townsin]. Grandeur of the Seas is the oldest ship (11 years) and had two previous dry dockings with an associated buildup of coatings and roughness (Table 5). Mariner of the Seas and Jewel of the Seas were 4 and 3 years old respectively. Both ships had no prior dry dockings, the outer coat was Intersmooth 460 SPC paint, and the hulls was in good condition (Table 6).

Table 5 Previous Paint applied to Grandeur of the Seas

PRODUCT	COLOR	COAT	DFT (microns)
Top Sides			
Intertuf 262	Grey	full coat	150
Intertuf 262	White	full coat	100
Interthane 990	White	full coat	50
Boottop			
Intertuf 262	Grey	Touch up	150
Intergard 267	Buff	touch up	100
Interspeed 640	Blue	full coat	75
Interspeed 640	Blue	full coat	75
Underwater Sides			
Intertuf 262	Grey	touch up	150
Intergard 267	Buff	touch up	100
Interspeed 640	Black	full coat	75
Interspeed 640	Red	full coat	75
Flats			
Intertuf 262	**Grey**	**touch up**	**150**
Intergard 267	Buff	touch up	100
Interspeed 640	Black	touch up	75
Interspeed 640	Red	full coat	75

Table 6 Existing Paint on Jewel of the Seas

PRODUCT	COLOR	COAT	DFT (microns)
Top Sides			
Intergard 269	Red	Full	50
Intershield 300	Aluminium	Full	200
Interthane 990	White	Full	50
Interthane 990	White	Full	50
Boottop			
Intergard 269	Red	Full	50
Intershield 300	Aluminium	Full	150
Intershield 300	Bronze	Full	150
Intergard 263	Light Grey	Full	50
Interthane 990	Cobalt Blue	Full	50
Interthane 990	Cobalt Blue	Full	50
Underwater Sides			
Intergard 269	Red	Full	50
Intershield 300	Aluminium	Full	150
Intershield 300	Bronze	Full	150
Intergard 263	Light Grey	Full	75
Intersmooth 465 SPC	Dark Red	Full	75
Intersmooth 460 SPC	Blue	Full	125
Intersmooth 460 SPC	Blue	Full	100
Flats			
Intergard 269	Red	Full	50
Intershield 300	Aluminium	Full	200
Intergard 263	Light Grey	Full	75
Intersmooth 465 SPC	Dark Brown	Full	120
Intersmooth 465 SPC	Dark Red	Full	120

4.2.2 Ship Operational Area

The geographical area in which the ships operate determines the presence and degree of fouling. Factors like temperature of the sea water, salinity etc. "A ship operating at slow speed in the Caribbean will foul at quite a different rate than a ship operating at high speeds in the North Atlantic which, in turn, will foul at a totally different rate than a ship sitting pierside in Norfolk". [Naval engineer's journal]. The operational regions for the Grandeur and Mariner were similar, however, the Jewel had been operating in European waters for 2 months prior to dry dock. The details are as follows.

1) The Grandeur of the Seas travel itinerary was mostly in Western Caribbean. This included Georgetown, Grand Cayman, Port Canaveral, Key West (Florida), Philipsburg, St Marten, St Thomas, Oranjestad, US Virgin Islands, Aruba, San Juan, Puerto Rico, Freeport, Grand Bahama Island, Cozumel, Mexico, and Belize City, Belize. From December 2005 to April 2006 round trip from Tampa to Cozumel plus Costa Maya. From April 2006 San Juan to Baltimore. From May until November 2006 round trip from Baltimore calling at Freeport, Bahamas; Key West and Port Canaveral, Florida; Cozumel and Costa Maya, Mexico - alternates with a five-night Bermuda itinerary. Round-trip from New Orleans for seven nights and round-trip from Baltimore for five nights. From May -November 2006 docking for two days at King's Wharf, Bermuda. Alternates with a nine-night Caribbean itinerary December 2006 to April 2007, to Cozumel, Grand Cayman and Costa Maya.

2) Mariner of the Seas travel itinerary was mostly in the Caribbean. This included round-trip from Port Canaveral, year-round on Sundays, alternating seven-night Eastern and Western Caribbean calling at Nassau, St. Thomas and St. Martin; or, Labadee, Ocho Rios, Grand Cayman, Cozumel Port Canaveral, Florida, Labadee, Hispaniola, Ocho Rios, Jamaica, Georgetown, Grand Cayman, and Cozumel, Mexico. It also included Port Canaveral, Florida, Coco Cay, Bahamas, Philipsburg, St Maarten.

3) Jewel of the Seas travel itinerary from January to May 2006 was between Eastern and Western Caribbean which includes Ft Lauderdale, Grand Cayman, Costa Maya, Key West, Cozumel, Sanjuan, St. Martin, St. Thomas, Cococay, and Antigua. From May to August 30, 2006, the ship traveled to Europe calling at Oslo, Copenhagen, Stockholm, Helsinki, St. Petersburg, and Tallinn. From June to July 25, 2006, she traveled between LeHarve, Plymouth, Cork, Dublin, Glasgow, and Norwegian Fjords including Bergen, Geiranger, Hellesylt, Flam, and Amsterdam. From September 11, 2006 she sailed round trip from Harwich. From Oct.7, 2006 she called at Martha's Vineyard, Massachusetts; Portland, Maine; Bar Harbor, Maine; Halifax, Nova Scotia; Saint John, New Brunswick. She completed a Baltic cruise (Harwich-Oslo-

Copenhagen-Stockholm-Helsinki-St Petersburg-Tallinn-Harwich) and in late October crossed the Atlantic to Martha's Vineyard, Massachusetts; Portland, Maine; Bar Harbor, Maine; Halifax, Nova Scotia, Boston. At the end of October, she left for Bermuda; San Juan, Puerto Rico; Charlotte Amalie, St. Thomas; Willemstad, Curacao; Oranjestad, Aruba. She spent from November 12, 2006 to April 21, 2007 in the Caribbean, calling at San Juan, St. Maarten, Antigua, St. Thomas, Nassau, Key West, Cozumel, Costa Maya, Grand Cayman.

4.2.3. Underwater Hull Cleaning Schedules

Underwater hull cleaning is used as a method of removing unwanted buildup of biofouling during the period between dry docking. The policy of some shipping companies is to perform propeller cleaning twice a year, and hull cleaning annually (according to availability dry docking, and other factors). The hull of the Grandeur had not been cleaned since the previous dry dock. Underwater cleaning of Mariner was performed on the following dates, April 2005, November 2005 and August 2006. The Jewel was probably cleaned in the spring of 2007; however, this has to be confirmed. None of the ships were reported to have been cleaned in the 6 months prior to dry dock and therefore the dry dock fouling condition represented the antifouling capabilities of the coating at that time.

4.3 Dry Dock Data

All three ships were dry docked during 2007. The dry docking occurred at three different locations (Table 7)

Table 7 Current Dry dock Schedule (2007)

Vessel	Most recent coating	Dry dock Dates	Days out of service	Time since last dry dock (months)	Dry dock location
Grandeur of the Seas	Nov-04	16-26 Apr 2007	10	30	Mobile
Mariner of the Seas	Oct-03	29 Apr to 6 May-2007	8	43	Bahamas
Jewel of the Seas	Apr-04	8-29 Oct-2007	21	42	Hamburg

The dry-docking data are presented with respect to the fouling condition, coating condition, surface preparation and new coating. At dry dock all three ships were painted with Intersmooth SPC and some selected areas with Intersleek. Section 4.7 has a detailed explanation of the advantages of using Intersleek on selected areas of the ships.

Grandeur of the Seas was painted with a sealer coat (aluminium) on the flats as the ship had tin based Interspeed 640 on its flats so as to comply with the IMO standards.

4.3.1 Fouling Condition

The three ships had different fouling communities (Table 8, Figure 24). Grandeur had the greatest fouling, followed by Mariner and Jewel. Grandeur had heavy algae and slime fouling attached to the hull, Mariner had light slime and Jewel of the Seas was very clean and only had light slime on boottop areas of the ship.

Grandeur had only been in service for 30 months since her last dry dock compared to over 40 months for Mariner and Jewel. Grandeur was coated with Interspeed 640 and had previous coatings build up from 2 maintenance dry docks compared to the other two ships that were new build and coated with Intersmooth 465 SPC. The greater fouling on Grandeur would therefore suggest that this coating was not performing as well as the Intersmooth 465 SPC.

Table 8 Assessment of the Average Fouling Condition for the Three Ships

	Slime %	Weed %	Barnacles %	Navy Rating
Grandeur				
Bow	68	75	0	20
Boottop	65	80	0	20
Side (P)	50	75	3	20
Side (S)	60	80	3	18
Flats	30	25	0	10
Keel	32	30	1	25
Mariner				
Bow	75	0	0	15
Boottop	15	0	0	0
Side (P)	85	0	2	16
Side (S)	90	0	3	18
Flats	50	0	0	5
Keel	25	0	1	2
Jewel				
Bow	48	0	0	10
Boottop	20	0	0	3
Side (P)	40	0	0	6
Side (S)	40	0	0	6
Flats	5	0	0	0
Keel	20	0	0	2

The differences in the fouling types that were present on specific areas of the hull were due to the condition of the antifouling coating, differences in the ambient conditions and underwater hull cleaning procedures. Algae fouling were typically more abundant at the bow, boottop and sides due to the presence of sunlight. Hard fouling in the form of encrusting bryozoans, barnacles and tubeworms were typically more abundant on the flats, keel and stern areas where there was reduced light and hydrodynamic shear stresses. The areas that were left unpainted during previous dry dockings due to docking blocks were more susceptible to fouling. These areas had the highest concentration of biofouling along the keel. Barnacle fouling was only found on the hulls of Grandeur and Mariner. All three sips, however, had barnacle fouling in the sea chests and hull openings.

The ship hull cleaning programs typically focus on the propellers, rudders, boottop and forward underwater sides of the hull. The flats and stern areas are usually not cleaned. The fouling communities were correspondingly less abundant in the areas where cleaning had occurred.

When the fouling was converted to the NSTM Rating it can be seen that propellers and rudders were the least fouled, followed by the forward part of the hull, with the flats and stern being most fouled. The standard error bars reflect the variation in fouling composition over a specific area.

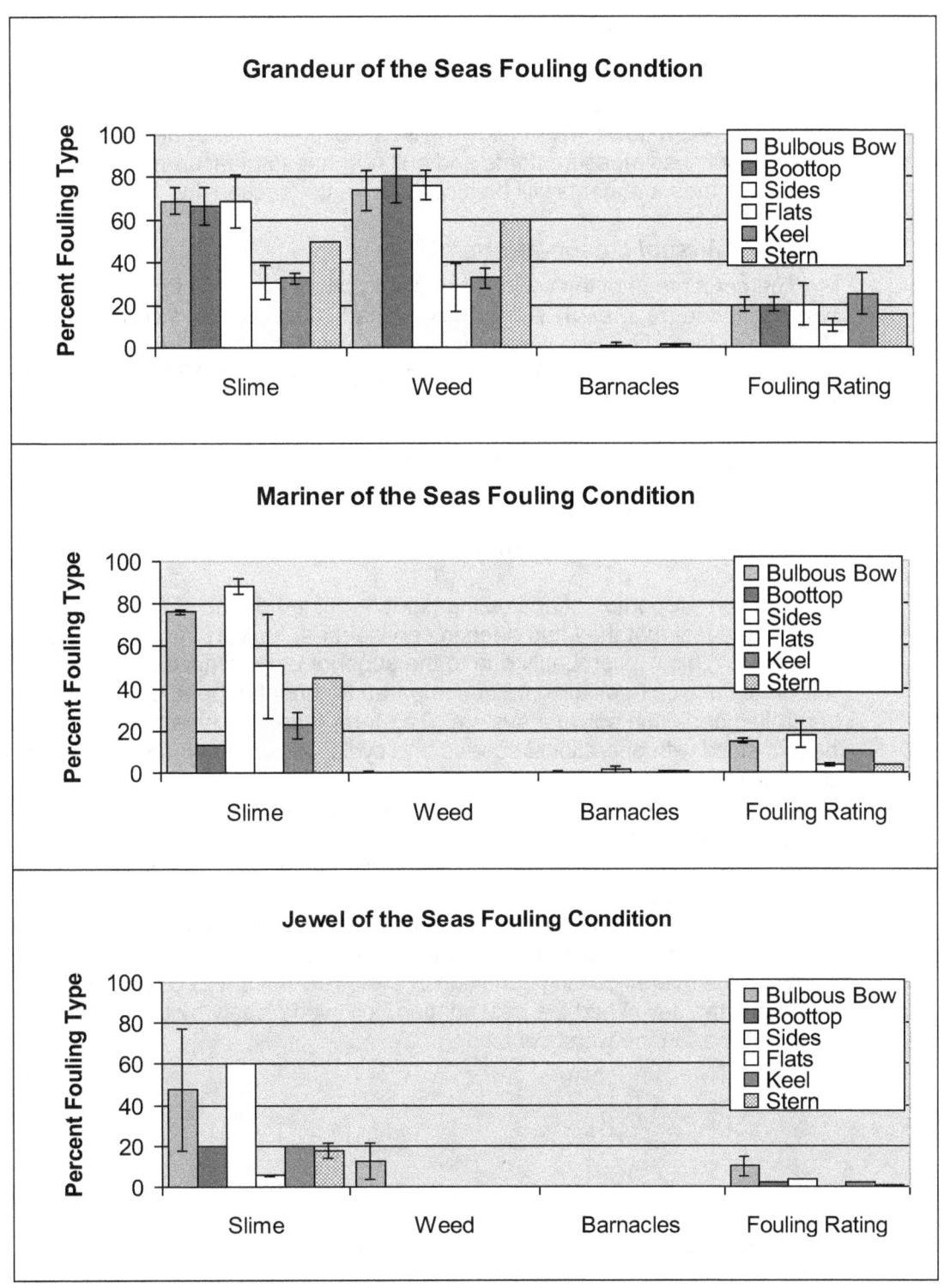

Figure 24 Comparison of Fouling for Three Ships

4.3.2 Coating Condition

The coating condition of each ship was observed after the high-pressure freshwater wash down. This included visual observations, photographic records, dry film thickness measurements and hull roughness measurements. The hull roughness measurements will be discussed in the section 4.6.

4.3.2.1 Visual Observations

The percentage of the coating remaining on the ship hull for each coating layer was estimated (Figure 25). All three ships had intact corrosion coats and the damage or loss of coating was most obvious with the antifouling coatings.

The condition of the antifouling coating is presented as AF1 (top antifouling coating) and AF2 (antifouling coating next to the corrosion coat). The color of these two coatings was different and therefore it was possible to differentiate when each coating was no longer present. Most of the antifouling on the bulbous bow of Grandeur was removed, either by anchor damage or from erosion of the coating. Antifouling on the sides was also depleted with the greatest amount of coating remaining on the flats and stern where hydrodynamic conditions were less aggressive.

A greater proportion of antifouling paint remained on the Mariner and Jewel despite the fact that they had been in service for at least 10 months longer than Grandeur. This was probably due to the superior performance of the Intersmooth 465 which is a self-polishing paint compared to the Interspeed 640 which is a controlled depletion polymer system. The former coating type typically performs best of ships with operational speeds and cycles similar to those of the ships discussed here.

Some blistering damage was seen for all the three ships. The Grandeur had the highest percent blister damage, followed by the Mariner and the Jewel had minimal blisters. The cause of the blisters may have been from interactions with solvents in the paint, poor surface preparation, inadequate coating thickness, and interactions with the cathodic protection systems. No definitive explanations could be made from data obtained during these surveys.

When the overall coating condition of the three ships was compared it was found that the Jewel had the best coating, followed closely by the Mariner, and that Grandeur was in the worst condition.

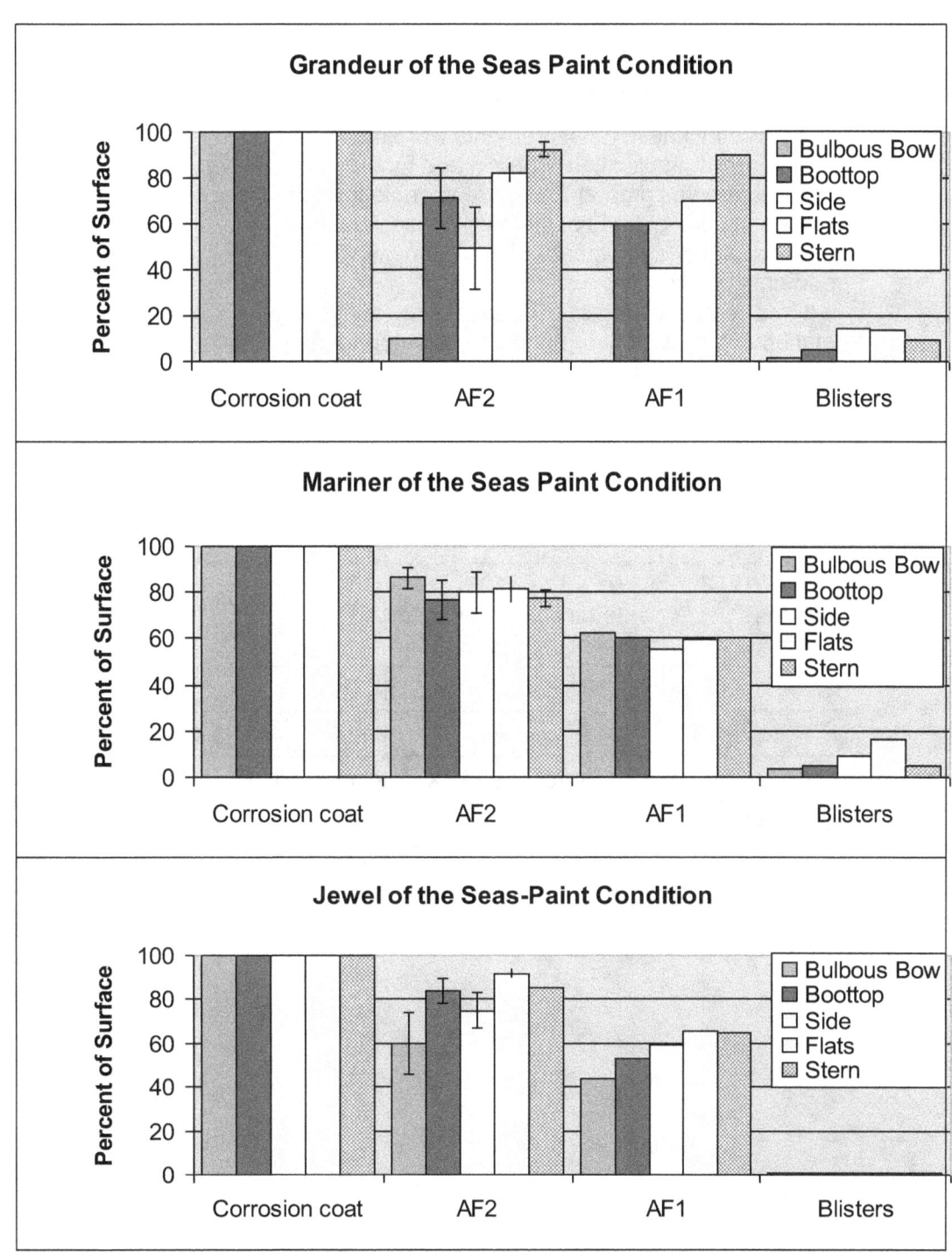

Figure 25 Comparison of Paint Condition for Three Ships

4.3.2.2 Dry Film Thickness

Dry film thickness measurements were made after high pressure wash down and after paint application (Table 9 and Figures 26 and 27). The coating thickness on Grandeur was greater and more variable than the other two ships. Both the Mariner and Jewel had coating thicknesses close to the initial coatings specifications and the standard deviations were fairly small indicating uniform and good paint application.

It has been shown that buildup of old coatings and DFT can lead to detachment, blistering damage and alligatoring effect. These all add to coating roughness and increased drag.

Table 9 Dry Film Thickness (DFT)

Ship	Grandeur of the Seas		Mariner of the Seas	Jewel of the Seas
After High Pressure Wash				
Average DFT (microns)	910 (intact paint)	426 (damaged paint)	711	727
SD	256	232	201	133
No Readings	94	96	114	173
After Dry Dock Coating Application				
Average DFT	890		875	947
SD	253		103	63
No Readings	251		148	315

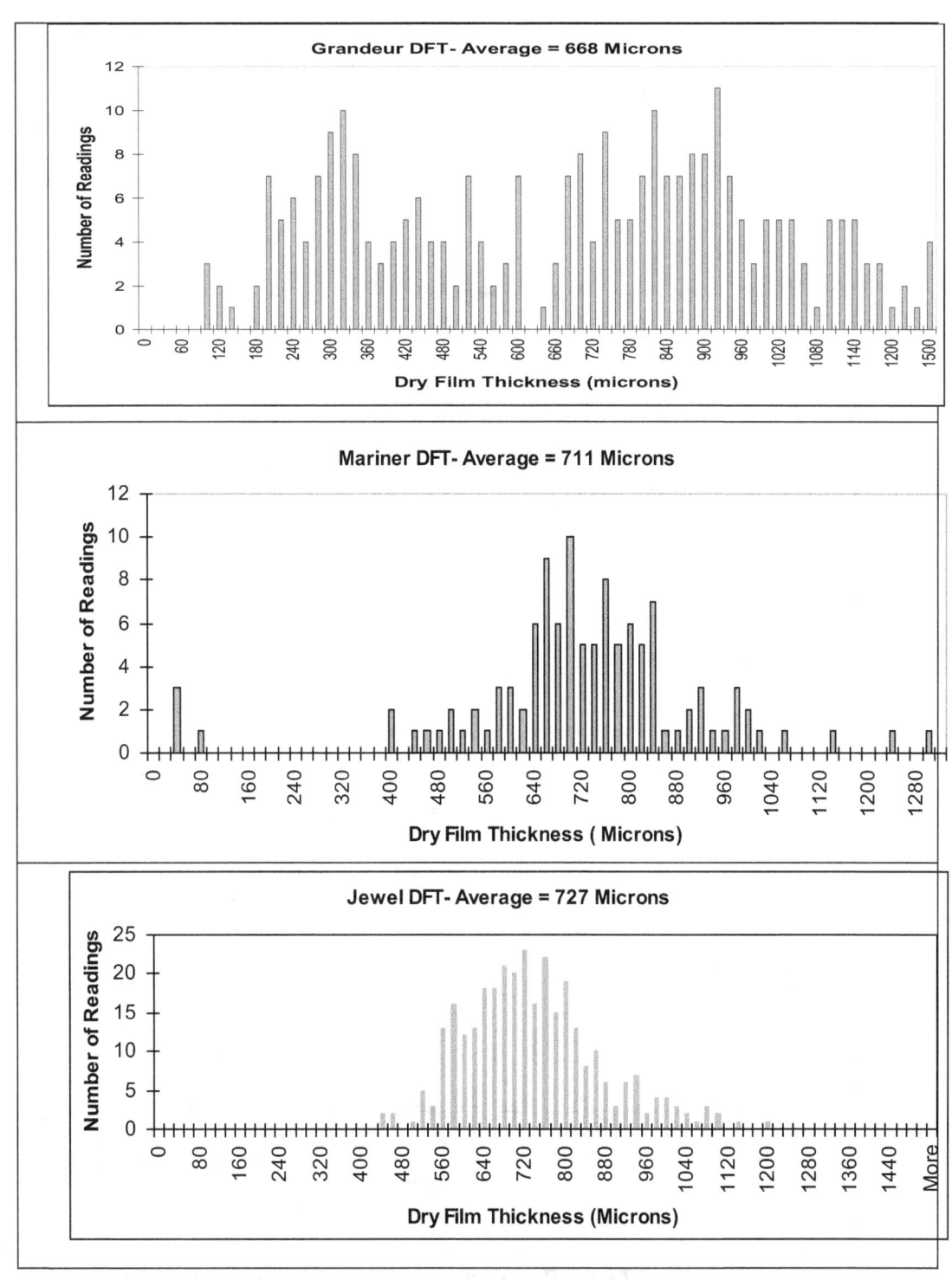

Figure 26 DFT Readings for the Three Ships after Washdown

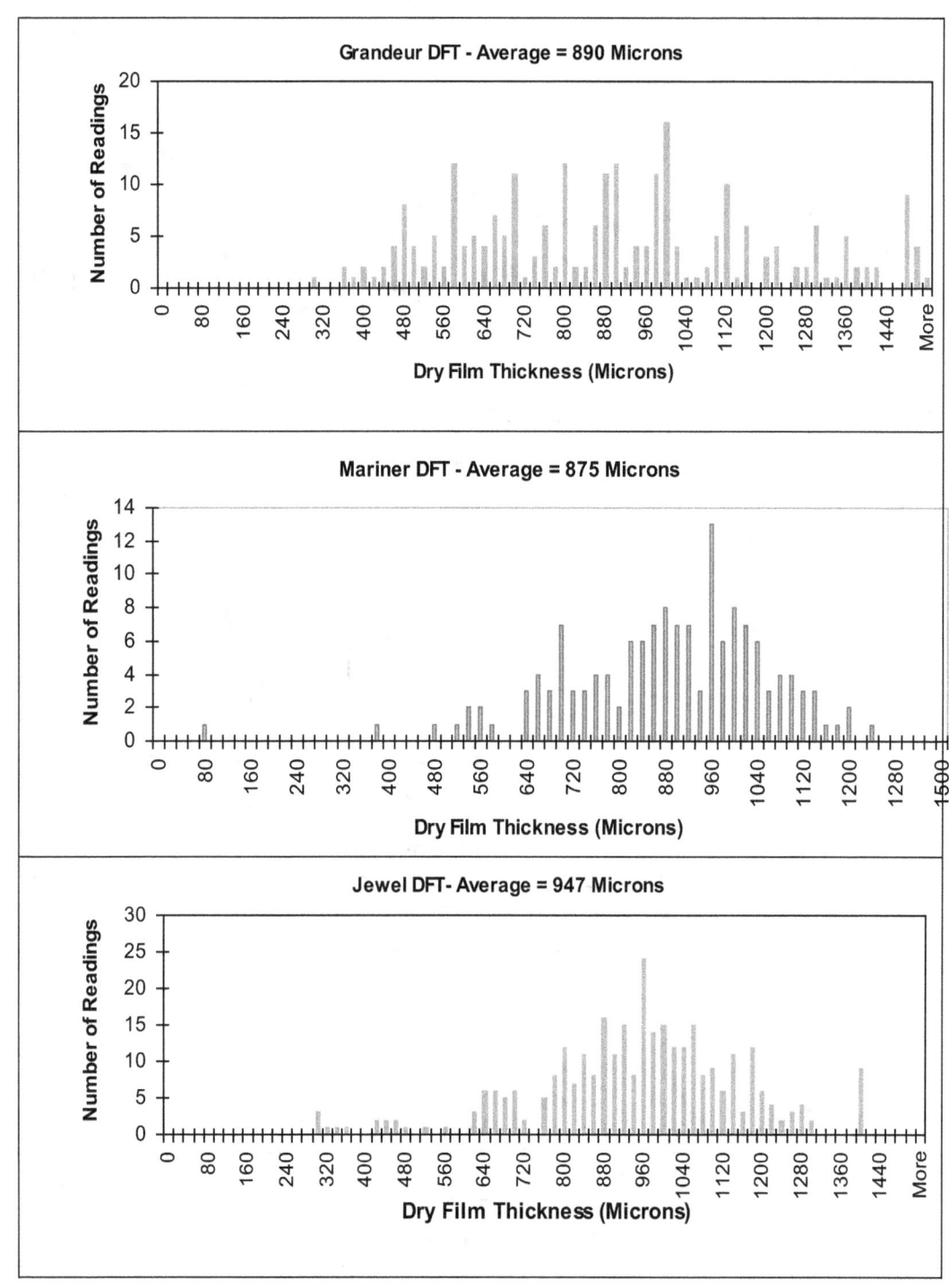

Figure 27 DFT Readings for the Three Ships after Coating

Grandeur of the Seas

The DFT readings for Grandeur after high pressure wash down showed a bimodal distribution. Where the coating was intact the average DFT was 910 microns, however, where the coating was damaged the average DFT was 426 microns. In several locations, the DFT exceeded 1500 microns which was the maximum thickness that could be measured by the Elcometer. (Figures 28 and 29)

Accumulation of paint from previous dry dockings had created a thick film. This caused the coating to detach which increased the roughness of the hull. Such paint accumulation is also referred to as "sandwich coatings" and, after fouling, represents the most serious cause of increased roughness. Studies from other ships have shown that the roughness due to peeling and flaking can increase average hull roughness by about 30 microns per year (Jotun).

In general, the following relationships were found between DFT and coating condition:

- If the average DFT was about 700-850 microns, the intact of paint was in good condition.
- If the average DFT was greater than 1500 microns, the paint was either detaching or blistering
- If the average DFT was less than 400 microns, the paint was either damaged or blistered. This may be because the paint was too thin to withstand the ICCP currents.

The variation in coating thickness was due to the age of the ship, number of dry dockings and number of paint layers.

After coating application, it was found that the average DFT had decreased. This was due to removal of coatings during a slurry blast clean which removed much of the damaged, blistered and delaminated paint.

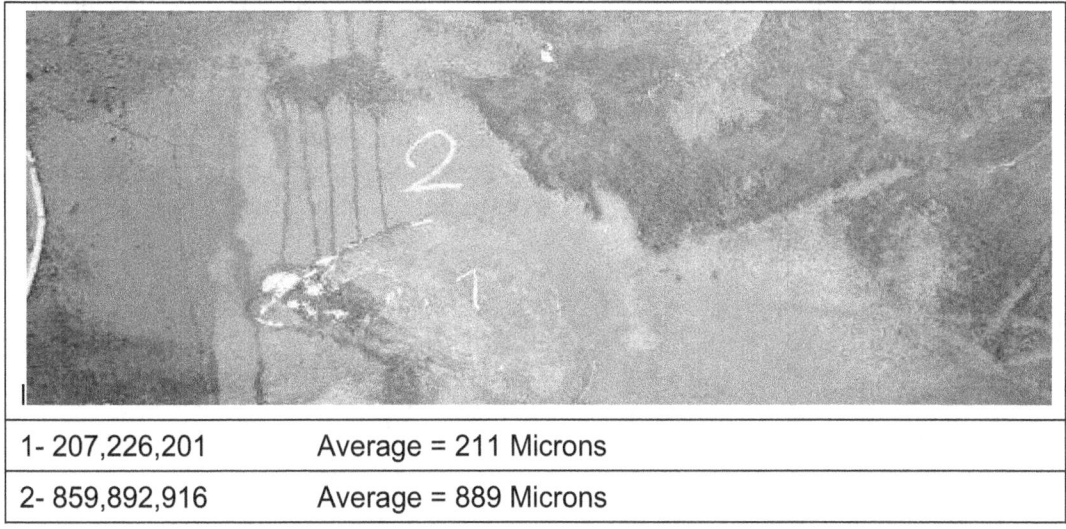

1- 207,226,201	Average = 211 Microns
2- 859,892,916	Average = 889 Microns

Figure 28 DFT Comparisons for Grandeur of the Seas

1-382,268,430	Average = 360 Microns
2-1600,1600,1600	Average = 1600 Microns

Figure 29 DFT Comparisons for Grandeur of the Seas

Mariner of the Seas

The average DFT for Mariner after high pressure wash down was 711 microns and the standard deviation was 201 microns. Unlike Grandeur the DFT thickness measurements displayed a more normal distribution, although the spread of coating thicknesses would suggest that there were disparities in the application of the paint coatings. For example, there was one area on the flats where the coating thickness exceeded 1500 microns DFT (Figures 30 and 31) and this led to delamination of the coating. Mariner has blisters throughout the length of the ship, figure 33 is a good example of the blisters from the flats on starboard side of the ship, these blisters were blasted in order to grind the area back to normal and to apply epoxy coat on all the grinded spots. After coating application, the average DFT increased by 160 microns to 875 microns. This increase in coating thickness was less than that specified in the new coating application.

Thickness is >1500 Microns

Figure 30 DFT Comparisons Mariner of the Seas

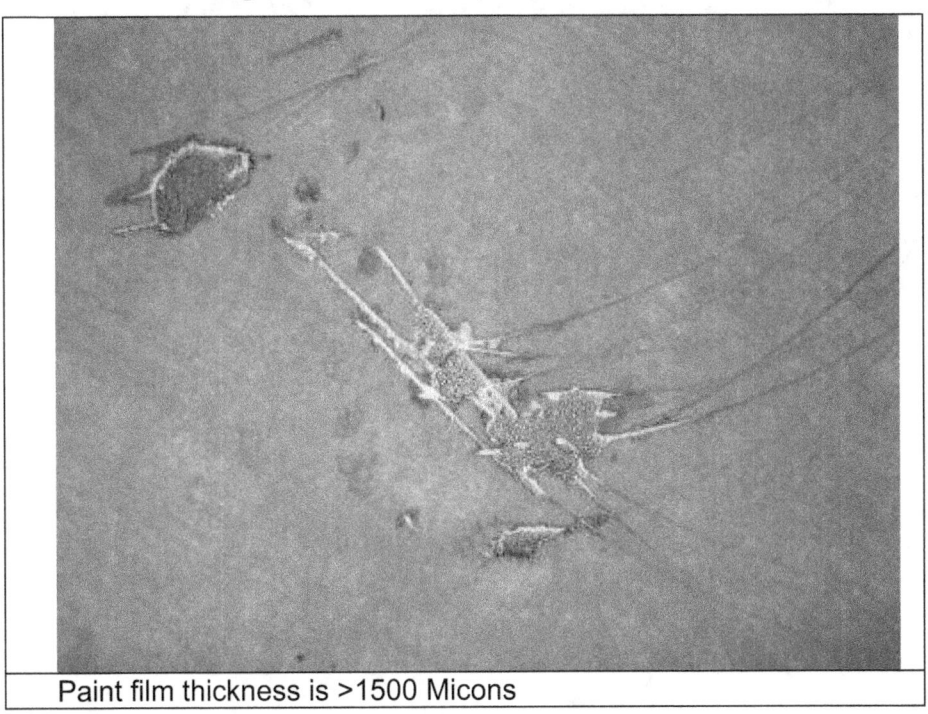

Paint film thickness is >1500 Micons

Figure 31 Mariner Paint Delamination of Coating on Flats with High DFT

Jewel of the Seas

The average DFT for Mariner after high pressure wash down was 727 microns and the standard deviation was 133 microns. The DFT thickness measurements displayed a normal distribution with only a small standard deviation which suggests that the coating was carefully applied (Figure 32). There were small areas of defects (Figure 33), however they were limited in their distribution and there was no apparent reason how they formed. After coating application, the average DFT increased by 220 microns to 947 microns. This increase in coating thickness was very close to that specified in the new coating application.

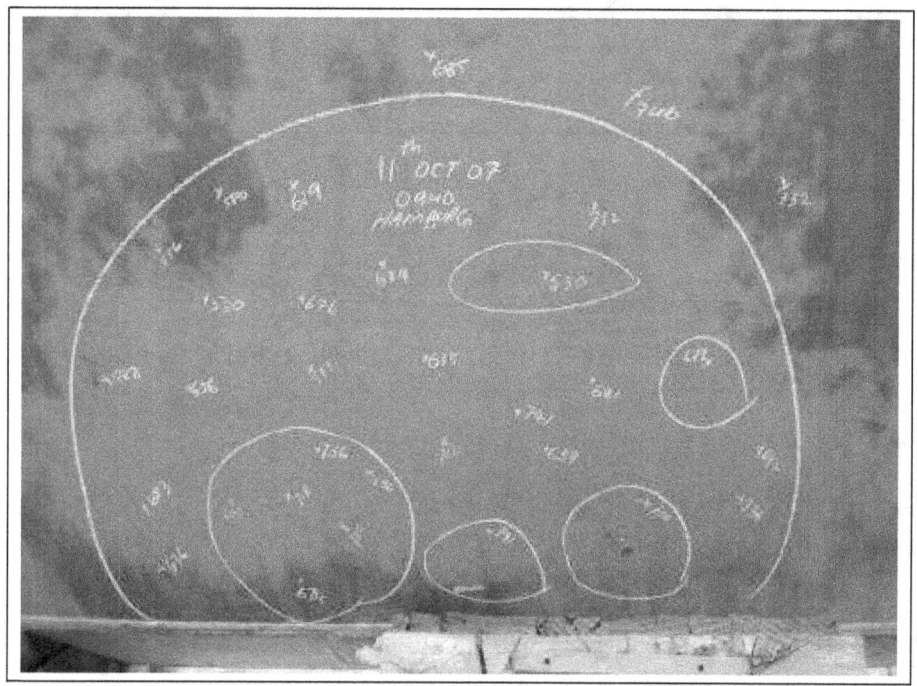

Figure 32 Jewel, Example of Uniform DFT

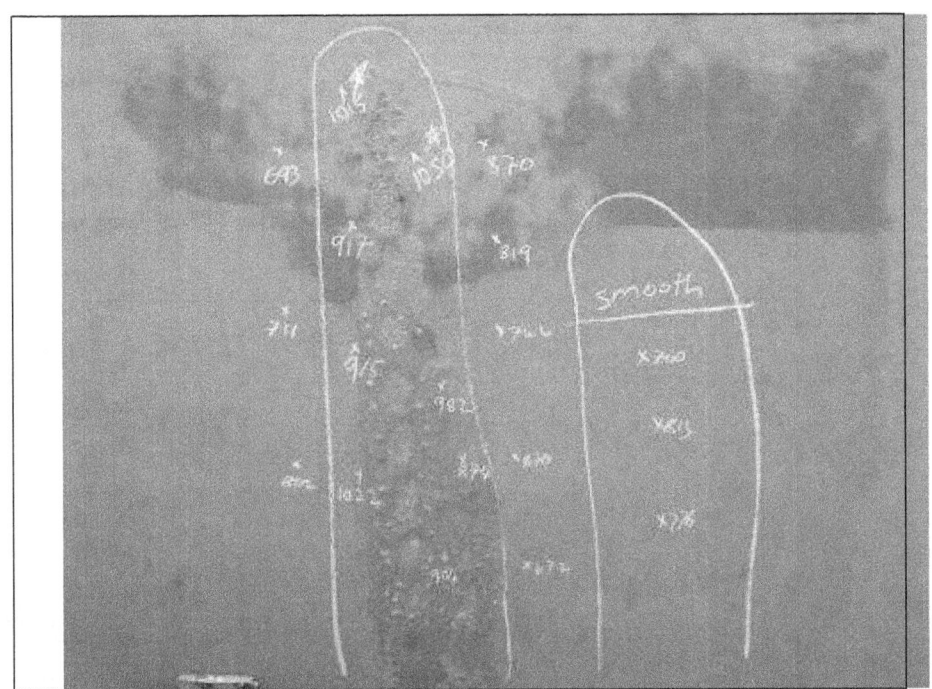

Figure 33 Example Delamination and Blisters near Bow Thrusters for Jewel

4.3.2.3 Took measurements (ASTM D4138-07)

Due to the high coating build up on Grandeur of the Seas a destructive Took coating test was applied. The measurements showed that there were 10 layers of paint on the hull (Figure 34). These multiple layers created roughness and coating instability. In this example the hull steel is the "bright silver" portion at the top left. The outermost antifouling is the red with some green slime film in the bottom right. Total estimated thickness was 2.02 mm (2020 microns)

Coating Destructive test using Proscope

Figure 34 Multiple Layers Seen Through the Proscope Microscope

Table 10 Thickness of Each Paint Layer

Layers #	DFT (mm)
1	0.375
2	0.202
3	0.289
4	0.289
5	0.231
6	0.115
7	0.115
8	0.115
9	0.231
10	0.058

Took measurements were not taken for Mariner and Jewel of the Seas as these ships did not have the multiple layers associated with previous dry dockings.

4.4 New Coating Application

This section describes the surface preparation and coating application for each ship. The underwater areas of the ship hull are divided into four general areas (Table 11 and Figure 35)

Table 11 Underwater Hull Areas for Painting

AREA	DESCRIPTION	Paint
Boottop (DLL to LLL)	The area which indicates the waterline with a painted line Draft load line to low load line.	Intersmooth 460 Tin Free SPC antifouling for 30 months (8,000 seamiles/month) in service period.
FWD-Side Bottom-Intersleek (LLL to Bilge Keel)	From the start of bow thrusters up till the end of bilge keel on the sides.	Intersleek 757 foul release coating for a 120 month in service period.
AFT-Side Bottom-Intersmooth 460 (LLL to Bilge Keel)	From the end of bilge keel on the sides till the stern	Intersmooth 460 Tin Free SPC antifouling for 30 months (8,000 seamiles/month) in service period.
Flat Bottom (Bilge Keel to Keel)	The flat surface on the bottom of a ship is flat bottom. Fins on the sides of a ship to reduce rolling of a ship is bilge keel, and keel is the back bone of the ship running from bow to stern for strength of a ship	Intersmooth 460 Tin Free SPC antifouling for 30 months (8,000 seamiles/month) in service period
Sea Chest and Hull Openings	Sea Chest is used to draw or discharge water, they are on the sides of the ship with both lower and upper sea chest openings.	Interswift 655 TBT-Free antifouling for 30-month in-service period.

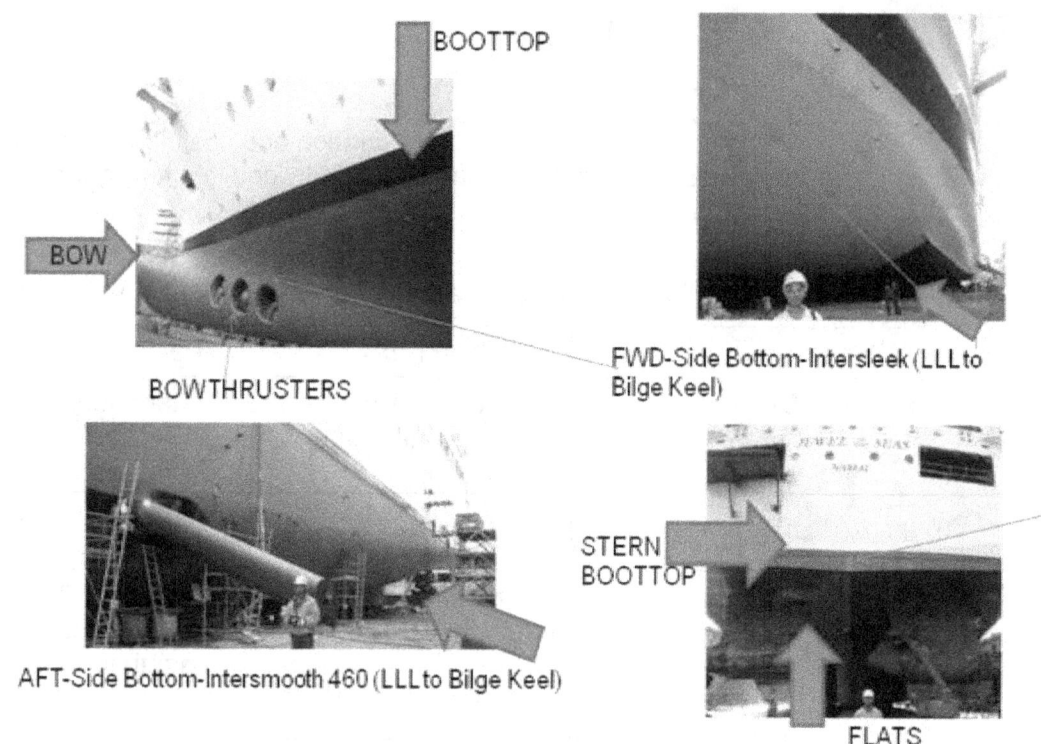

Figure 35 Underwater Areas for Painting

4.4.1 Grandeur of the Seas

What happened in the dry dock

As soon as the vessel was dry docked and secured, the hull surface was subjected to a high pressure wash (>5000 psi). Marine growth, old antifouling and loose paint were removed followed by a slurry sweep blast on the boottop, vertical sides and flats. The anode dielectric shields were sand blasted to sa 2.5. Priority was given to rudders, thruster openings and stabilizer areas in order to commence work on these units. All sacrificial anodes, ICCP anodes, ICCP reference cells, logs, echo sounders, sliding blocks, rubber gaskets, propeller blades and shafts were covered in order to protect them from surface preparation and coatings application.

An epoxy sealer coat was applied to the flats over the self-polishing copolymer tributyltin paint. Two coats of epoxy were applied over the entire hull, followed by two coats of SPC copper on the flats and the boottop, and three coats of SPC copper on the remainder of the ship. The high build epoxy anode dielectric shield was applied using a trowel. Coating application continued at night.

Table 12 Paint Applied in the Dry dock for Grandeur of the Seas

PRODUCT	COLOR	COAT	DFT (microns)	VOLUME (US Gallons)
BOOTOP				
Intertuf 262	red	full	150	46.5
Intergard 264	black	full	100	28.3
Intersmooth 460 SPC	blue	full	125	142.0
SIDE BOTTOM				
Intertuf 262	red	full	150	348.5
Intergard 264	black	full	100	212.4
Intersmooth 460 SPC	dark red	full	100	849.8
Intersmooth 460 SPC	dark brown	full	100	424.9
FLAT BOTTOM				
Intertuf 262	red	10% touch up	150	31.0
Intergard 264	black	10% touch up	100	18.9
Intertuf 203	aluminium	full coat	75	283.2
Intersmooth 460 SPC	dark brown	full coat	125	473.0
Intersmooth 460 SPC	dark red	full coat	125	473.0
CROSSOVER TANKS				
Intertuf 262	red	10% touch up	150	5.2
Intergard 264	black	10% touch up	100	3.2
Interspeed 640	red	full coat	100	81.8
Interspeed 640	black	full coat	100	40.9
ANODE SHIELDS				
Intergard 822	red	full coat	400	42.2

How the ship left the dry dock

The ship had favorable weather conditions and enough time to complete all the painting jobs as per the specifications before it left the dry dock. In general, the coating buildup was high and thickness exceeded the recommended values. This will encourage paint film detachment.

4.4.2 Mariner of the Seas

What happened in the dry dock

As soon as the vessel was dry docked and secured, the hull surface was subjected to a high-pressure wash (>5000 psi). Priority was given to rudders, thruster openings and stabilizer areas to finish first in order to commence work on these units. All sacrificial anodes, ICCP anodes, ICCP reference cells, logs, echo sounders, sliding blocks, rubber gaskets, propeller blades and shafts were covered in order to protect them from surface preparation and coatings application. The damaged areas mostly on the boottop were spot ground. After grinding the damaged spots, light high-pressure freshwater washing was applied to the hull. Epoxy touchup coats were applied to the damaged areas. Two coats of SPC tin free antifouling were then applied to the flats and the fouling release coating Intersleek was applied to the vertical sides. Coating continued through the night.

Table 13 Paint Applied to Mariner of the Seas

PRODUCT	COLOR	COAT	DFT (Microns)	VOLUME (US Gallons)
BOOTOP				
Intertuf 262	red	full	150	47
Intergard 264	black	full	100	30
Intersmooth 462 SPC	blue	full	125	90
FWD SIDE BOTTOM				
Intergard 264	red	5% touch up	125	13.2
Intersleek 717	mid brown	full	100	314.5
Intersleek 737	light pink	full	100	269.1
Intersleek 757	grey	full	150	353.6
AFT SIDE BOTTOM				
Intertuf 262	red	5% touch up	150	12.4
Intergard 264	black	5% touch up	100	7.6
Intersmooth 460 SPC	dark brown	full	135	410.1

Intersmooth 460 SPC	dark red	full	135	410.1
FLAT BOTTOM				
Intertuf 262	red	5% touch up	150	17.8
Intergard 264	black	5% touch up	100	10.9
Intersmooth 460 SPC	dark brown	full coat	110	476.1
Intersmooth 460 SPC	dark red	full coat	110	476.1
CROSSOVER TANK				
Intertuf 262	red	10% touch up	150	7.7
Intergard 264	black	10% touch up	100	4.7
Interspeed 640	red	full coat	125	76.3
TOPSIDES				
Interthane 990	white	20% touch up	100	30

How the ship left the dry dock

This vessel had a very short time in dry dock to complete the painting. There was insufficient time for a full cure of the final coat of dark red antifouling 460 SPC on the first bow thruster pocket and the vertical side and the hull markings with Interthane 990.

4.4.3 Jewel of the Seas

What happened in the dry dock

As soon as the vessel was dry docked and secured, the hull surface was subjected to a high-pressure wash (>5000 psi). All sacrificial anodes, ICCP anodes, ICCP reference cells, logs, echo sounders, sliding blocks, rubber gaskets, propeller blades and shafts were covered in order to protect them from surface preparation and coatings application. Damaged areas were ground to bare steel and touched up with three coats of epoxy. A full coat of Intersmooth SPC antifouling was then applied to aft-side bottom and flats. A full coat of Intersleek 900 (red) was applied to the forward sides and bottom. The sea chests and hull openings were coated with Interswift 655 TBT free system.

Table 14 Paint applied in the dry dock for Jewel of the Seas

PRODUCT	COLOR	COAT	DFT (microns)	VOLUME (US Gallons)
BOOTOP				
Intershield 300	aluminum	5% touch up	200	5.1
Intergard 263	light grey	5% touch up	75	1.98
Intersmooth 460 SPC	blue	Full coat	100	86.6
FWD SIDE BOTTOM				
Intershield 300	aluminum	2% touch up	150	7.55
Intersleek 717	mid brown	full coat	100	279.5
Intersleek 737	pink	full coat	100	264.8
Intersleek 757	grey	full coat	150	314.5
AFT SIDE BOTTOM				
Intershield 300	aluminum	2% touch up	200	4.28
Intergard 263	light grey	2% touch up	75	1.69
Intersmooth 460 SPC	dark brown	2% touch up	100	3.19
Intersmooth 460 SPC	dark red	full coat	135	217
FLAT BOTTOM				
Intershield 300	aluminum	2% touch up	200	10
Intergard 263	light grey	2% touch up	75	4
Intersmooth 460 SPC	dark brown	2% touch up	100	7.5
Intersmooth 460 SPC	dark red	full coat	110	414
SEACHEST				
Intershield 300	aluminium	10% touch up	200	6.9
Intergard 263	light grey	10% touch up	75	2.7
Interswift 655	brown	full coat	100	33.4
Interswift 655	dark red	full coat	100	36

How the ship left the dry dock

The ship had favorable weather conditions and enough time to complete all the surface preparation and coatings applications as per specifications. When it left the dry dock, the coating was in excellent condition.

4.5 Cost of Paint

This section provides an estimate of paint costs for the different areas of each ship.

4.5.1 Paint Cost Calculations

The costs of paint applied to each ship were calculated using the cost in dollars per liter of paint, recommended DFT, and surface area to be covered. The cost of paint varies and is negotiable, however, representative prices were used for these calculations and are shown in Table 15

Representative Cost of Paint

Table 15 Approximate Estimation of Paint Cost

Type of Paint	$/Liter
Epoxy coat	9.2
Antifouling Hybrid	13.2
Sealer coat	5.0
Self-Polishing Copolymer Antifouling	15.6
Intersleek Link coat	11.9
Intersleek Tie coat	47
Intersleek Top coat	47

Example calculation for the costs for Intersleek 717 link coat to the forward side bottom area of the Jewel of the Sea. The Intersleek is a three-part system which includes a link coat (717), tie coat (737) and finish coat (757)

Cost Intersleek Link Coat	$11.9/liter
Recommended DFT	100 microns
% solids	54
% loss factor	40
Surface area	4000 m^2

The wet film thickness required for the specified DFT is calculated as follows:

WFT= (DFT x 100/%solids)

WFT= (100 x 100/54) = 185 microns

The volume of paint (liters) required is calculated as follows:

Liters = (WFT(μm)/1,000,000(μ/m)) x area (m²) x [(100+%Loss)/100] x 1,000l/m³

Liters = (185/1,000,000) x 4000 x (100+40)/100 x 1,000 = 1037

The total cost of paint is calculated as follows:

$/liter x total number of liters

 = $11.9 x 1,037 = $12,340.00

Similar calculations were performed for each area of each ship and the results are shown in Table 16. The paint coverage rate is entered using the paint specifications. The full calculations are presented for the Jewel of the Seas and summary findings for Grandeur and Mariner. (Table 17)

Jewel of the Seas Surface area calculations
Boottop (800 m²)
Forward side bottom (4000 m²)
Aft side bottom (1700 m²)
Flat bottom (4000 m²)
Sea chest and hull openings (550 m²)

Table 16 Paint Calculations for Jewel of the Seas

Boottop (DLL to LLL) [800 m²]

Product	% Loss Factor	% Solids	DFT Microns	$/Liter	Liters Used	Total price USD
Intershield 300	40	60	200	9.2	18.65	172
Intergard 263	40	57	75	7.1	7.35	52
Intersmooth 460 SPC	40	40	100	19	280	5320
			485			5544

FWD-Side Bottom Intersleek (LLL to Bilge keel) [4000 m²]

Product	% Loss	% Solids	DFT Microns	$/Liter	Liters Used	Total price USD

Intershield 300	40	60	150	9.2	28	258
Intersleek 717	40	54	100	11.9	1037	12340
Intersleek 737	40	57	100	47	982	46154
Intersleek 757	40	74	150	47	757	35579
						94331

Aft Side Bottom Intersmooth 460 (LLL to Bilge Keel) [1700 m^2]

Product	% Loss	% Solids	DFT Microns	$/Liter	Liters Used	Total price USD
Intershield 300	40	60	200	9.2	15.9	146
Intergard 263	40	57	75	7.1	6.26	44
Intersmooth 460 SPC	40	40	100	15.6	11.9	186
Intersmooth 460 SPC	40	40	135	15.6	803	12527
						12903

Flat Bottom (Bilge Keel to Keel) [4000 m^2]

Product	% Loss	% Solids	DFT Microns	$/Liter	Liters Used	Total price USD
Intershield 300	40	60	200	9.2	37.34	344
Intergard 263	40	57	75	7.1	14.74	105
Intersmooth 460 SPC	40	40	100	15.6	28	437
Intersmooth 460 SPC	40	40	110	15.6	1540	24024

			485			24910

Seachest and Hull openings [550 m²]

Product	% Loss	% Solids	DFT Microns	$/Liter	Liters Used	Total price USD
Intershield 300	40	60	200	9.2	25.7	236
Intergard 263	40	57	75	7.1	10.1	72
Interswift 655	40	58	100	13.7	133	1822
Interswift 655	40	58	100	13.7	133	1822
			475			3952

Thinners

Product	Volume liters	Total price USD
International	65	$950

Two types of antifouling paints Self Polishing Copolymer 460 copper and Intersleek were taken for the calculations.

There is always 30-40% of loss factor taken in to account while painting.
Cost of SPC paint for square meter (m²) = $15.6/liter
Cost of Intersleek paint for square meter (m²) = $105.9/Liter (includes link, tie and finish coats)
So, the cost of the paint is a key factor for the ship owners to choose the type of antifouling paint for their ships. The above cost is just for the paint and the labor costs, dry dock expenses are not included.
Ultimately to get the maximum benefit from the paints and to cut down the costs some of the important high efficiency areas are painted with Intersleek antifouling and the rest of the ship with SPC 460 antifouling. Section 4.7 has the detailed explanation. Bow and the bow thruster area mostly gets mechanical damage while at sea when the anchor is used moreover the Intersleek is sensitive to the mechanical damage so there is an advantage of using SPC 460 in bow area

Table 17 Three Ships Estimated Paint Costs

	Area coated	Length	Type of Antifouling	Volume in Liters	Cost USD
Grandeur	Boottop (DLL to LLL)	600 m²	Intertuf 262	173	684
			Intergard 264	105	750
			Intersmooth 460 SPC	526	8182
	Side Bottom (LLL to Bilge Keel)	4500 m²	Intertuf 262	1295	5130
			Intergard 264	788	5617
			Intersmooth 460 SPC	4725	73650
	Flat Bottom (Bilge Keel to Keel	4000 m²	Intertuf 262	115.1	460
			Intergard 264	70	497
			Intertuf 203	1050	5270
			Intersmooth 460 SPC	3500	54557
	Crossover Tanks	672 m²	Intertuf 262	22.8	91
			Intergard 264	11.8	84
			Interspeed 640	456	6013
	Anode Shields	220 m²	Intergard 822	248	392
	TOTAL	9992 m²		15063	**161377**
Mariner	FWD-Side Bottom-Intersleek (LLL to Bilge Keel)	4500 m²	Intergard 264	49.2	349
			Intersleek 717	1167	13870
			Intersleek 731	1105	51978
			Intersleek	1277	60055

71

			757		
	AFT-Side Bottom (DLL to Bilge Keel)	3200 m²	Intertuf 262	46.05	184
			Intergard 264	28	199
			Intersmooth 460 SPC	2912	45390
	Flat Bottom (Bilge Keel to Keel	4600 m²	Intertuf 262	66.15	265
			Intergard 264	40.25	286
			Intersmooth 460 SPC	3542	55212
	Crossover tank	1000 m²	Intertuf 262	28.8	115
			Intergard 264	17.5	124
			Interspeed 640	282	2983
	TOTAL	13500 m²		15341	**231010**
Jewel	Boottop (DLL to LLL)	800 m²	Intershield 300	18.65	172
			Intergard 263	7.35	52
			Intersmooth 460 SPC	280	5326
	FWD-Side Bottom-Intersleek (LLL to Bilge Keel)	4000 m²	Intershield 300	28	258
			Intersleek 717	1037	12329
			Intersleek 737	982	46202
			Intersleek 757	757	35588
	AFT-Side Bottom-Intersmooth 460 (LLL to Bilge Keel	1700 m²	Intershield 300	15.86	146
			Intergard 263	6.26	45
			Intersmooth	1398	21795

		460 SPC		
Flat Bottom (Bilge Keel to Keel	4000 m²	Intershield 300	37.34	344
		Intergard 263	14.74	105
		Intersmooth 460 SPC	1940	45828
Seachest and Hull Openings	550 m²	Intershield 300	25.7	236
		Intergard 263	10.1	72
		Interswift 655	266	3647
TOTAL	11050 m²		12648	**172145**

4.6 Hull Roughness

It is estimated that for each 20 microns (10^{-3} mm) increase in hull roughness there will be a 1 to 2 per cent increase in fuel consumption [Swain 2007]. Approximately 70-80% of the total shaft horsepower of a large ship is utilized to overcome friction with the water. Hence, "the maximum possible smoothness is of fundamental importance for economic operation" [Jotun 1985]. Increases in roughness due to peeling, flaking, blistering and general deterioration has been shown to increase average hull roughness by about 30 microns per year.

The roughness of a newly applied coating is also important. Roughness will be affected by factors such as surface preparation, weather conditions, painter's skills, equipment and dry dock quality control. It has also been found that for a good quality SPC coating any roughness created during dry dock will be reduced once the ship is in service and the coating polishes to a smoother condition.

Hull roughness measurements were taken on the three cruise ships before and after paint application (Table 18). The Grandeur had significantly greater roughness than the other two ships.

Table 18 Hull Roughness

Pre D/D Condition	Grandeur of the Seas	Mariner of the Seas	Jewel of the Seas
Average roughness	264	115	77
SD	208	118	96
Number of Readings	88	46	202
Post D/D Condition			
Average roughness	253	103	63
SD	164	39	70
Number of Readings	47	43	175

4.6.1 Hull Roughness for Grandeur of the Seas

The Grandeur of the Seas was built in 1996 and had 2 prior dry dockings. The ship had never been completely blasted back to bare steel and therefore there was a significant buildup of old coatings. The ship active antifouling coating was Interspeed 640 which is a hybrid paint and doesn't polish as well as the SPC systems. The hull was roughness was associated with coating deterioration, the presence of blisters and an inferior antifouling system for this type of vessel. The average roughness of this ship was 264 microns which might be expected to cause a 10 to 20% penalty in performance (Figure 36).

The in-docking average roughness was 264 microns and the out-docking average roughness was 253 microns. The standard deviation, however were extremely high and demonstrate the variability in the coating roughness over the whole ship.

The comparison of power and fuel consumption for Grandeur propulsion tells us that prior to dry dock the fuel consumption was about 30% higher than the new ship condition. This was due to slime and weed fouling, coating damage and high hull roughness. after dry dock the fuel consumption was about 18% higher than the new ship condition. This was due to macro and micro roughness left after the new coating application. For the months of August and September 2007 (4 months after dry dock); the fuel consumption had dropped to about 13% higher than the new ship condition. This may be due to polishing of the paint and reduced micro roughness.

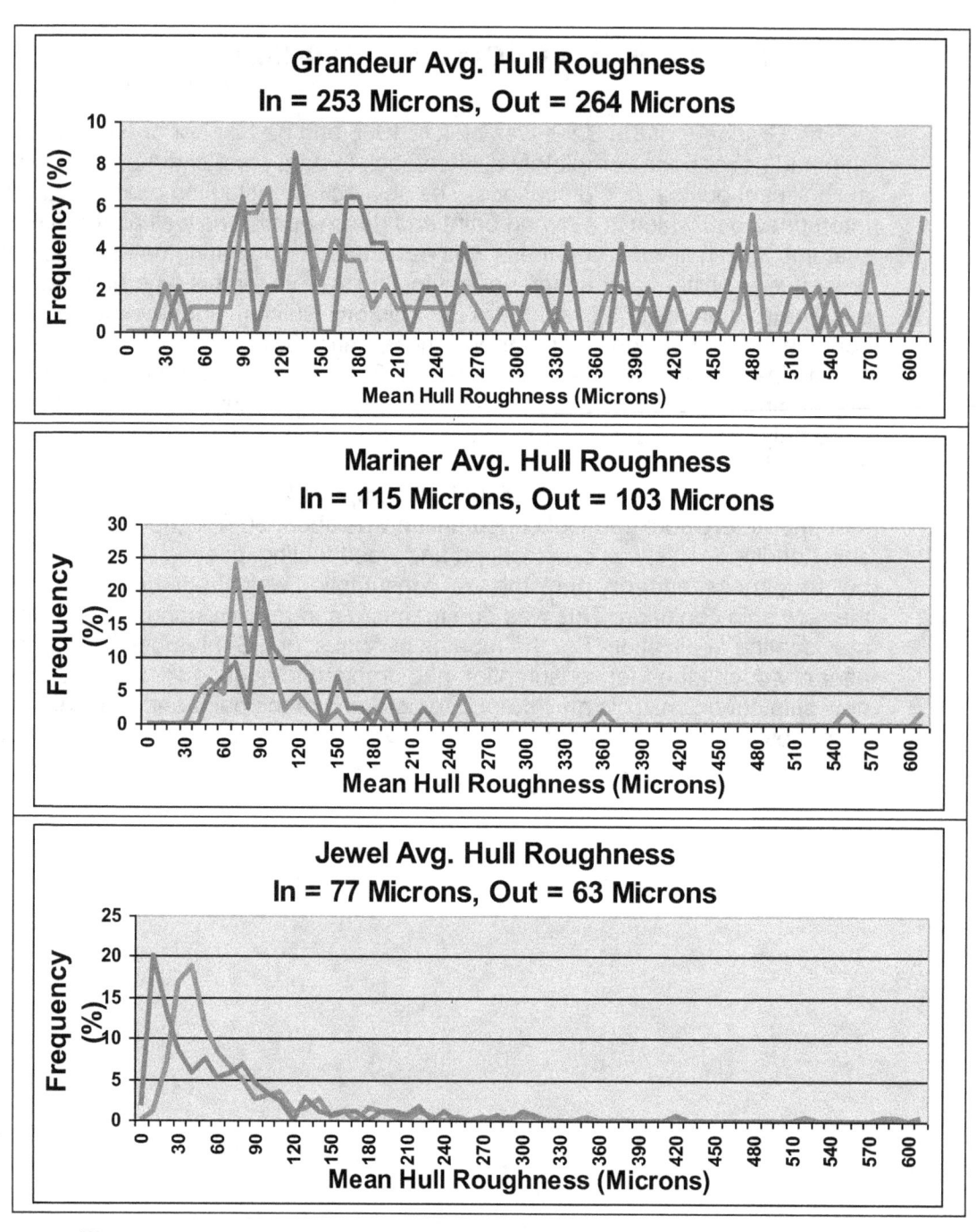

Figure 36 Indocking and Outdocking Roughness for the Three Ships

4.6.2 Hull Roughness for Mariner of the Seas

The Mariner of the Seas was built in 2003 and had no prior dry dockings. The active antifouling coating was the Intersmooth 460 SPC system. The average hull roughness of the coating for in-docking was 115 microns and the out-docking were 103 microns (Figure 36). The standard deviation for the in-docking was high and reflects the roughness at some damaged areas of the hull. The standard deviation decreased after the coating application. There is no significant difference between the in-docking and out-docking values.

4.6.3 Hull Roughness for Jewel of the Seas

The Jewel of the Seas was built in early 2004 and had no prior dry dockings. The active antifouling coating was the Intersmooth 460 SPC system. The average hull roughness of the coating for in-docking was 77 microns and the out-docking were 63 microns (Figure 36). The excellent coating condition at out-docking reflected the high standards at this yard, good weather conditions and plenty of time in the drydock schedule. The ship was very clean and smooth.

4.6.4 Hull Roughness for SPC Copper and Intersleek Paints

During the dry dock for Jewel hull roughness measurements were made of the newly applied Intersmooth 460 SPC and the Intersleek paints (Figure 37). The average hull roughness for the Intersmooth was 90 microns and greater than the Intersleek which was 70 microns Table 19.

Table 19 Average Hull Roughness Comparison of Intersmooth and Intersleek

	Intersmooth 460 SPC	Intersleek
Average roughness	90	70
SD	10	10
Number of Readings	30	30

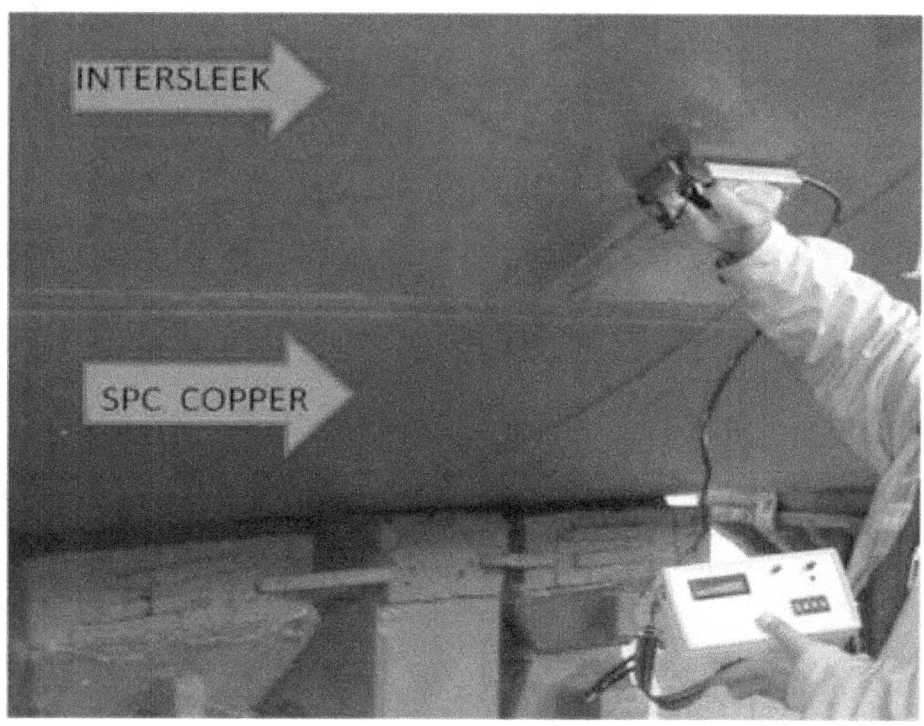

Figure 37 Comparison for SPC Copper and Intersleek Roughness

4.7 Coating Selection.

It is a good strategy to apply both SPC copper and Intersleek paint to the ships. This is based upon the requirements of different parts of the ship for hydrodynamic performance, protection from mechanical damage and cost. For example, for the Jewel, the hull was divided into high efficiency and high loading areas.

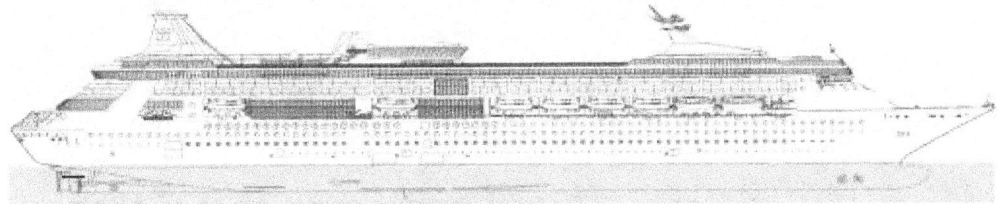

279.6

Figure 38 Intersleek Paint Area for Jewel of the Seas

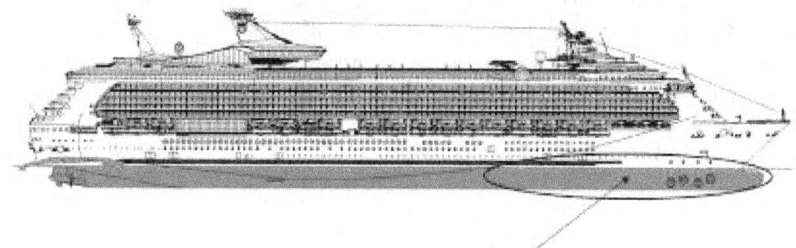

CRITICAL HYDRODYNAMIC AREAS OF A SHIP

High efficiency area coating system:

Figure 39 Critical Hydrodynamic Areas of a Ship

4.7.1 High Efficiency Areas: This area is responsible for approximately 50% of the overall resistance (Kari Pihlajaniemi) and, therefore, requires the greatest attention with respect to hull roughness (Figure 40). This area was painted with Intersleek, which has a lower hull roughness than the SPC system. It is, however, more expensive and vulnerable to mechanical damage.

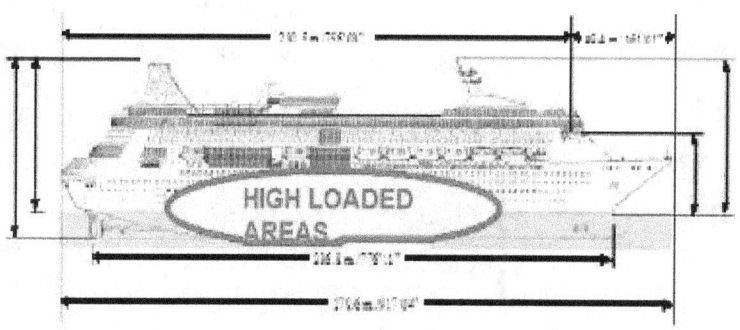

HIGH LOADED AREAS

Figure 40 High Loaded Areas on a Ship

4.7.2 High Loaded Areas: The two areas of the ship most vulnerable to mechanical damage are the bulbous bow and boottop. The bulbous bow area is subject to high erosion from sea water flow and sediments (when present), mechanical damage from flotsam and jetsam, and mechanical damage from the anchor chains. The coatings applied to the boottop area of the hull are subjected to wet and dry cycles, oil, and other pollutants that may be present at the air/seawater interface and also mechanical damage from floating debris in the

80

water and fendering. Both these areas require a stronger coating and are typically coated with a SPC system.

CHAPTER V

CONCLUSIONS

This research investigated the condition of the underwater hull areas of three ships at three different dry dockings. The ships represented three different classes, ship builders, ages, operating schedules coating systems and dry docks. The data has provided a unique opportunity to identify the possible causes in differences in hull coating performance.

Analysis of the data is divided into observations made before coating, surface preparation and paint application, and after coating application and has enabled the following conclusions to be reached:

Before Coating
1. The age of the ship affects outer hull coating condition when accumulations of prior coatings are allowed to develop. This can only be prevented by periodic blasting to bare steel and recoating. Grandeur had two prior dry dockings and coating build up and was in poor condition. Mariner and Jewel still had their original coating and were in good condition.
2. Grandeur of the Seas had the greatest fouling, followed by Mariner of the Seas and Jewel of the Seas
3. Grandeur was painted with the less expensive hybrid copper-based antifouling and Mariner and Jewel were coated with more advanced copper SPC systems. The copper SPC systems are more suitable for ships with this service schedule.
4. The paint condition for Grandeur was poor with delamination and blisters. This was due to minimum surface preparation and maintenance at previous dry dockings.
5. The paint condition for Mariner and Jewel was good.
6. The average dry film thickness for Grandeur was high and variable. This was due to paint film build up at prior dry dockings.
7. The average DFT for both Mariner and Jewel were good (about 700 microns). Mariner had greater variability in DFT than Jewel which suggests that the paint application for Jewel was better.
8. The DFT measurements lead to the following overall conclusions. If the paint is too thick, there are increased chances of coating damage and blistering. Blistering may also be a problem when the paint is too thin.

When the paint is uniformly applied according to the paint supplier specifications, then the paints perform well and have a very smooth surface

9. The hull roughness for Grandeur was high. This reflected lack of maintenance and surface preparation from prior dry dockings and the use of a hybrid antifouling system.
10. The hull roughness analyses for Mariner and Jewel showed they had smooth hulls; Jewel was better than Mariner.
11. Analyses of powering and fuel consumption for Grandeur demonstrated that the poor hull condition had a negative impact on the fuel consumption.

surface preparation and paint application

1. The surface preparation is important to the subsequent success of the new coating application.
2. The Grandeur only had high pressure wash and slurry blast to remove loose coating. The ship really required a full blast back to bare steel.
3. The Mariner and Jewel had a high-pressure wash and the damaged areas were ground back. This was the correct decision for these coatings.
4. Grandeur had a sealer coat, epoxy and antifouling systems added to the remaining coatings. It was difficult to achieve a smooth surface.
5. Mariner and Jewel had touch up epoxy coatings over damaged areas and two coats of SPC copper applied to all but the forward sides where Intersleek was applied.
6. The Mariner had insufficient dry dock time for the coating system to be completed. The ship was refloated before the required drying time for some of the areas.
7. The Jewel was successfully recoated.
8. The overall conclusions for surface preparation and coating application are that the success of the new coating application requires correct surface preparation, sufficient dry dock time to complete the job, and quality control.

After Coating

1) DFT measurements after coating application demonstrated:
    a. Grandeur had a large variability in DFT.
    b. The change in DFT after coating application for the Mariner showed that the increased in DFT was less than theoretically should have been applied which suggests losses during spraying.
    c. The change in DFT after coating for the Jewel showed that increase in DFT closely matched the theory and quality control was good.
12. Hull roughness of the new coatings varied for each ship. All ships were coated with the same SPC copper system and therefore these differences reflected prior surface preparation and coating application.

a. Grandeur had the highest roughness and was most likely associated with inadequate surface preparation.
b. Jewel had the lowest hull roughness and was associated with excellent dry dock practice.

Attention to the ship hull condition with respect to hull roughness and biofouling is a key factor in maintaining efficiency and reducing fuel costs. This was demonstrated by analyses of the Grandeur Voyage Data before and after dry dock. The following summarizes the findings.

- prior to dry dock the fuel consumption was about 30% higher than the new ship conditions. This was due to slime and weed fouling, coating damage and high hull roughness.
- after dry dock the fuel consumption was about 18% higher than the new ship condition. This was due to macro and micro roughness left after the new coating application.
- for the months of August and September 2007 (4 months after dry dock) the fuel consumption had dropped to about 13% higher than the new ship condition. This may be due to polishing of the paint and reduced micro roughness.

REFERENCES

Abbott A, Abel PD, Arnold DW, Milne A. Cost-Benefit Analysis of the use of TBT. The Case for a Treatment Approach. Sci. Total. Environ. 258:5-19. year 2000.

American Bureau of shipping (ABS) Guide for the Class Notation, Coating Performance Standard (CPS) December 2006

American Bureau of Shipping, USA. Guidance notes on The Inspection, Maintenance and Application of Marine Coating Systems page 56, year 2007

Antifouling Product and Technology Guide, International Paint ltd, Akzo Nobel, 2004

Butterworth-Heinemann, Ship Construction 2001, Fourth Edition,

Connie G Clavier (cost of ownership) Naval Ship Self-Assessment of Hull Powering Performance Using Propulsion Shaft Torsionmeters and GPS, Naval Engineers Journal May 1997, page182

Champ M A.2003. Economic and Environmental Impacts on Ports and Harbors from the Convention to Ban Harmful Marine Antifouling Systems. Mar. Pollut. Bull. 46:935-940.

Development of Standard Procedures for Shipboard Print Inspection U.S. Department of Transportation Maritime Administration in Cooperation with National Steel and Shipbuilding Company San Diego, California, April,1988 NSRP 0295

General Council of British Shipping. Recommended Practice for the Protection and Painting of Ships British Maritime Technology Limited,

Improving the Field Reliability of Protective Coatings, Proceedings of the 6th Technical Symposium of the Steel Structures Painting Council 2-5 November, 1987.Pittsburgh, PA 15213

International Maritime Organization TBT

International Paint, Inc. USA, Antifouling The Legislative Position Key Points Summary (update) International Coatings Ltd UK, February, 2004

Jelic-Mrcelic G, Sliskovic M, Antolic B. Biofouling Communities on Test Panels Coated with TBT and TBT-free copper-based Antifouling Paints. Biofouling 22:293-302.year 2006.

Jotun Marine Coatings, Division of Jotun A/S, paint manufacturers, Sandefjord, Norway, 1985

Kirk William Schumacher, An Instrumented Rotating Brush Device to Evaluate the Removal of Biofouling from Non-Toxic, Antifouling Coatings.

Kari Pihlajaniemi, the draft proposal for the GN coating strategy for RCL, April 6th 2007.

Michael P. Schultz Effects of Coating Roughness and Biofouling on Ship Resistance and Powering, Department of Naval Architecture and Ocean engineering, United States Naval Academy, Annapolis, Maryland USA 2007.

National Association of Corrosion Engineers, Coatings & Linings Handbook 1985.

Naval Ship's Technical Manual Chapter 081. Waterborne Underwater Hull cleaning of Navy ships, Revision 5. pages (81-1), APR 2002.

Royal Caribbean International, Press Center, Press Release. 2007

Swain Geoffrey, Brett Kovach, Arthur Touzot, Franck Casse, Christopher J Kavanagh. Measuring the performance of today's antifouling coatings. Journal of Ship Production, Aug 2007, V23, n. 3, pp. 164-171

Swain G W, Griffith J R, Bultman J D, Vincent H L. "The Use of Barnacle Adhesion Measurements for the Field Evaluation of Non-Toxic Foul Release Surfaces." Biofouling, Vol. 6, pp. 105-114, 1992.

Swain G W, Schultz M P. "The Testing and Evaluation of Non-Toxic Antifouling Coatings." Biofouling, Vol. 10.

Townsin

US Congress "Organotin Antifouling Paint Control Act of 1988"

U.S. Navy, Performance Testing of Marine Coatings: New Test and Evaluation Procedures, U.S. Department of Transportation Maritime Administration and the U.S. Navy. June, 1990, NSRP 0311

Wallsend Research Station, Wallsend, 1986

Yule A B, Walker G. "Temporary Adhesion of the Barnacle Cyprid: The Existence of an Antennular Adhesive Secretion." J Mar Biol Assoc, Vol. 64, pp. 679-686, 1984

Cover Photo credit to Matthew Barra from Pexels.